ALSO BY PAUL A. OFFIT, MD

Overkill: When Modern Medicine Goes Too Far

*Bad Advice: Or Why Celebrities, Politicians, and Activists
Aren't Your Best Source of Health Information*

*Pandora's Lab:
Seven Stories of Science Gone Wrong*

*Bad Faith: When Religious Belief
Undermines Modern Medicine*

*Do You Believe In Magic?:
The Sense and Nonsense of Alternative Medicine*

*Deadly Choices:
How the Anti-Vaccine Movement Threatens Us All*

*Autism's False Prophets:
Bad Science, Risky Medicine, and the Search for a Cure*

*Vaccinated: One Man's Quest to Defeat
the World's Deadliest Diseases*

*The Cutter Incident: How America's First Polio Vaccine
Led to the Growing Vaccine Crisis*

YOU BET YOUR LIFE

From Blood Transfusions to Mass Vaccination,
the Long and Risky History of Medical Innovations

PAUL A. OFFIT, MD

BASIC BOOKS

New York

Cover design by Ann Kirchner
Cover images © Natal'â Maâk/EyeEm via Getty Images; © nexusby/Shutterstock.com;
© m2art/Shutterstock.com; © Martial Red/Shutterstock.com
Cover copyright © 2021 by Hachette Book Group, Inc.

Basic Books
Hachette Book Group
1290 Avenue of the Americas, New York, NY 10104
www.basicbooks.com

Printed in the United States of America.

First Edition: September 2021

Published by Basic Books, an imprint of Perseus Books, LLC, a subsidiary of
Hachette Book Group, Inc. The Basic Books name and logo is a trademark of
the Hachette Book Group.

The Hachette Speakers Bureau provides a wide range of authors for speaking events.
To find out more, go to www.hachettespeakersbureau.com or call (866) 376-6591.

The publisher is not responsible for websites (or their content) that are not owned
by the publisher.

Print book interior design by Jeff Williams.

Library of Congress Cataloging-in-Publication Data
Names: Offit, Paul A., author.
Title: You bet your life : from blood transfusions to mass vaccination, the long and
    risky history of medical innovation / Paul A. Offit.
Description: First edition. | New York : Basic Books, 2021. | Includes bibliographical
    references and index.
Identifiers: LCCN 2020054280 | ISBN 9781541620391 (hardcover) | ISBN
    9781541620384 (ebook)
Subjects: MESH: Therapies, Investigational—history | Diffusion of Innovation | Drug
    Therapy | Risk Assessment | Bioethical Issues | Biomedical Research—history
Classification: LCC RM300 | NLM WB 300 | DDC 615.5/8—dc23
LC record available at https://lccn.loc.gov/2020054280

ISBNs: 9781541620391 (hardcover), 9781541620384 (ebook)

LSC-C

Printing 1, 2021

"The past is never dead. It's not even past."

—WILLIAM FAULKNER, *Requiem for a Nun*

CONTENTS

YOU BET YOUR LIFE

WE'RE AT THE DAWN OF A WONDROUS AGE: WE CAN REPROGRAM OUR immune systems to attack cancers that until now had been a death sentence, like those of the brain, pancreas, and lung. We can genetically modify animals (such as pigs) to offer an endless supply of hearts for transplant, eliminating the waiting list for thousands of people. We are close to vaccines to prevent Alzheimer's disease, Parkinson's disease, and dementia. Viruses that kill bacteria can be injected into people whose infections are resistant to all antibiotics. Artificial blood can eliminate our reliance on the blood of strangers and its occasional contamination with viruses, known or unknown. A gene-editing system called CRISPR will allow us to reengineer human genes, dramatically lessening the horror of single-gene diseases, such as cystic fibrosis and sickle cell disease.

There is, however, one catch. And it's not a small one. As you'll see in the pages that follow, virtually every medical breakthrough has exacted a human price. Some might argue that the solution is simple: Just wait until new technologies have been broken in. Wait until the learning curve is over. Those who need a heart transplant, for example, could just wait until all the kinks are worked out before being one of

1

the first to receive a pig's heart. But among the four thousand people currently on the transplant list, about 1,300 will die while waiting. Either way, both those who wait and those who choose a pig's heart are gambling.

I first started working on this book at the beginning of one of the worst pandemics in history. On November 17, 2019, a bat coronavirus made its debut in the human population. The virus was called SARS-CoV-2 and the disease was called COVID-19 (coronavirus disease-2019). Just one year after it had arrived, the virus had infected hundreds of millions of people and killed more than a million, mostly from pneumonia. And it was just getting started.

Although this bat coronavirus was new, human coronaviruses have been around for decades. First identified in the early 1960s, human coronaviruses cause coughs, colds, sore throats, and pneumonia. At our hospital, the Children's Hospital of Philadelphia, human coronaviruses account for about 20 percent of all respiratory infections during the winter. This bat virus, however, was different.

- SARS-CoV-2 was "the angel of death" in nursing homes, accounting for more than 40 percent of all COVID-19 deaths. For influenza virus, which kills about thirty thousand people every year, nursing homes account for fewer than 10 percent of deaths.

- SARS-CoV-2 raged during the summer months. No one had anticipated this. Human coronaviruses— as well as other respiratory viruses, such as influenza, that are spread by small droplets in coughs and sneezes—are winter diseases. They always disappear during the summer.

- SARS-CoV-2 caused some people to lose their sense of taste or smell for weeks at a time. Scientists have now found that the virus can enter the brain through nerves in the nose. Human coronaviruses don't do this. No respiratory viruses do this.

- SARS-CoV-2 caused an unusual, multi-organ disease in children that mimicked another disease called Kawasaki's (a relatively rare disorder characterized by rash, pink eye, cracked lips, swollen glands, a "strawberry" tongue, swelling of the hands and feet, peeling of the skin, and occasionally fatal inflammation of the arteries that supply blood to the heart). SARS-CoV-2 also mimicked a disease called toxic shock syndrome, which is as bad as it sounds. No other virus does this to children.

- SARS-CoV-2 caused inflammation of the lining of blood vessels (vasculitis), which, in addition to causing liver and kidney disease, increases the risk of strokes and heart attacks. This phenomenon was made all the more amazing by the observation that SARS-CoV-2 doesn't enter the bloodstream. Rather, the virus induces the body's own immune system to destroy cells that line veins and arteries. Human coronaviruses don't do any of these things either.

- Some people who caught and survived SARS-CoV-2 never made an immune response against the virus. This suggested that the virus might be suppressing the immune system, like the AIDS virus. This unusual form of immune suppression was especially true in men, who were twice as likely to die from the disease.

All of these surprises occurred within the first year of the virus's debut. It is likely that more surprises lie ahead.

With hundreds of thousands of people dying from COVID-19, everyone was desperate for a cure. In April 2020, I was selected by Dr. Francis Collins, director of the National Institutes of Health (NIH), to participate on a committee to accelerate the development of vaccines and treatments. I was also a member of the Food and Drug Administration (FDA) vaccine advisory committee. These positions gave me a front-row seat for how people were about to make decisions under uncertainty. Again, there were no risk-free choices. And desperation led to bad decisions.

On April 7, 2020, the FDA approved an antimalarial drug called hydroxychloroquine for the treatment of COVID-19, even though no study had ever found that it worked. Some people, like the president of the United States, reasoned that at the very least it was worth a try. What could be the harm? Unfortunately, any drug that has a positive effect can also have a negative effect. During the next few weeks, several studies showed that hydroxychloroquine neither treated nor prevented COVID-19. Worse, about 10 percent of recipients suffered severe heart arrhythmias, some fatally. Two months later, the FDA withdrew its recommendation. The most disturbing aspect of the hydroxychloroquine disaster was that the drug hadn't gone through the usual process of FDA licensure, which takes about a year of careful, thorough review. Rather, it had been approved under something called emergency use authorization—a faster, less thorough system that would soon be applied to the approval of coronavirus vaccines.

Although much disagreement surrounded the best way to treat COVID-19, everyone agreed that the only way to stop its spread was with a vaccine. As a consequence, never in history had more money or more expertise been devoted to making a single medical product. The World Health Organization (WHO), the Bill and Melinda Gates Foundation, and the US Department of Health and Human Services poured tens of billions of dollars into the effort. By the end of 2020, more than one hundred companies were pushing forward with a vaccine. Every strategy that had ever been used to make a vaccine was employed to prevent the spread of SARS-CoV-2, as well as several that had never been tried before. Some of these novel vaccines were being mass-produced in a program called Warp Speed before anyone really knew whether they worked or were safe.

In other words, by the end of 2020, an elusive, difficult-to-characterize bat coronavirus that had surprised doctors with its novel symptoms and unusual pathologies was soon to be countered by a series of vaccines that had never before been used against any other virus in history.

This book is about capturing the emotions of that moment at the end of 2020, when phrases like "Warp Speed," "the race for a vaccine," and "vaccine finalists" scared people that corners were being cut, time lines compressed, and safety measures ignored. It shouldn't be surprising that a CNN poll in the summer of 2020—when a thousand people were dying every day from the coronavirus—found that less than half of all Americans were willing to take a COVID-19 vaccine if offered. But again, a choice not to get a coronavirus vaccine was not a risk-free choice; rather, it was a choice to risk infection, hospitalization, long-term disabilities, and possibly death.

In the pages that follow, we'll examine the triumphs and tragedies behind nine of modern medicine's greatest advances: transplants, blood transfusions, anesthesia, biologicals, antibiotics, vaccines, X-rays, chemotherapies, and genetic engineering. All of these stories provide poignant lessons about when and whether to accept new technologies.

During the twentieth century, the life span of Americans has increased by thirty years. The medical breakthroughs profiled in this book have been directly responsible for much of that increase. Unfortunately, all of these breakthroughs have been accompanied by tragedy. And if we don't pay attention to these painful lessons from the past, we'll be condemned to repeat them.

PART I

RISK

BY THE END OF 2020, SEVERAL FORCES WERE AT PLAY: IN THE UNITED States alone, SARS-CoV-2 was killing one thousand people every day. Several companies had launched large-scale tests of their coronavirus vaccines, during which tens of thousands of people were injected with either the vaccine or saltwater (placebo) to determine whether the vaccines worked and were safe. Most of the strategies used to make these vaccines had never been used before. They were chosen because they were the easiest to construct and the fastest to mass-produce, not because they were necessarily going to be the best. Dr. Anthony Fauci, head of the National Institute of Allergy and Infectious Diseases at NIH, predicted that a coronavirus vaccine could be available by the end of the year, certainly no later than early 2021. Typically, it takes fifteen to twenty years to develop a vaccine, but, if Fauci was right, these COVID-19 vaccines would be developed in a year. By spending more than $20 billion on large clinical tests and mass production, the US government had taken the risk out of making vaccines for pharmaceutical companies, dramatically compressing time lines.

Additionally, none of these new vaccines would be subjected to the typical licensing procedures required by the FDA, which usually take about a year. Rather, they would be approved through emergency use authorization, which would allow vaccines to be injected into the arms of Americans within moments of rolling off the assembly lines. People would soon be forced to decide between the risk of getting COVID-19 and the risk of getting a vaccine that had not been subjected to the typical research, development, testing, and licensure processes.

Americans making these decisions can't have been comforted by the past. As has been true for virtually every vaccine ever made, the first vaccines aren't always the best, safest, and last. For example, a live, weakened polio vaccine introduced in 1963 was replaced by an inactivated polio vaccine in 2000, when it became clear that the former actually *caused* polio in eight to ten US children every year. The first measles vaccine in 1963—which caused a high rate of fever and rash—was replaced by a safer, better vaccine in 1968. Another measles vaccine, which was also introduced in 1963, was taken off the market when it was found to actually increase the risk of pneumonia. The first rubella (German measles) vaccine in 1969, which caused arthritis in small joints like fingers and wrists, was replaced by a safer vaccine in 1979. The *Haemophilus influenzae* type b (Hib) vaccine in 1985, a bacterial vaccine that wasn't particularly effective in young children, was replaced by a far more effective one in 1987. And the first shingles vaccine in 2011, designed to prevent one of the most debilitatingly painful diseases, was replaced by a much better one in 2017.

In the next three chapters, we'll examine the human stories behind heart transplants, blood transfusions, and anesthesia. In each instance, we'll also discuss when during the evolution of these technologies it was worth taking the risk and when it wasn't. Although it's always easy to judge in retrospect, lessons can be found in each of these stories that might shed light on similar decisions today.

CHAPTER 1

LOUIS WASHKANSKY

Heart Transplants

A CHIMERA, ACCORDING TO GREEK MYTHOLOGY, HAS A LION'S HEAD, A goat's body, and a serpent's tail. The Minotaur, who roamed the island of Crete, was half human and half bull. Centaurs were half man, half horse. We are at once fascinated and horrified by mythical beasts that combine one animal part with another. The world's first heart transplant evoked the same reaction—for the same reason.

——

Boyd Rush was in bad shape. For years, the sixty-eight-year-old retired upholsterer, who was deaf and mute, had lived alone in the Laurel Trailer Park on the outskirts of Jackson, Mississippi. High blood pressure had taken a toll on his heart, which was failing. As a result of inadequate circulation, his left leg was black with gangrene and his face was dotted with blood clots. On January 21, 1964, after suffering another heart attack, Rush was taken to the University of Mississippi Medical Center. Comatose and with a faint pulse, doctors inserted a breathing tube into his windpipe and attached it to a mechanical ventilator. On January 22, one day after he arrived at the hospital, his left leg was amputated. (Boyd Rush would never be a candidate for a heart transplant today.)

Dr. James Hardy, a heart surgeon at the medical center, had been waiting for this moment. A native of Newala, Alabama, and a graduate of the University of Pennsylvania School of Medicine, Hardy was the first person in the United States to successfully transplant a human lung. After having transplanted hearts into more than two hundred experimental animals, he was ready to perform the first human heart transplant in history. The problem was finding a donor. "At the outset," said Hardy, "it was expected that months, or perhaps even years, might elapse before an acceptable donor and recipient died simultaneously."

Hardy knew that a trauma victim was in the intensive-care unit of his hospital. He also knew that, although this patient was brain dead, his heart was still beating. Hardy's only option would have been to remove life support and wait for the heart to stop. Believing this was unethical, he refused to do it. (Four years would pass before the diagnosis of brain death allowed doctors to turn off breathing machines in patients whose hearts were still beating.) "Since we were not willing to stop the ventilator," said Hardy, "we had concluded that a situation might arise in which the only heart available for transplantation would be that of a lower primate." Hardy had prepared a large chimpanzee named Bino in the operating room next to Rush's. At the time, the sale of monkeys and chimps in the United States was unregulated.

On January 23, 1964, James Hardy sutured Bino's heart into Boyd Rush's chest. After shocking it with a defibrillator, Bino's heart jumped back to life. Unfortunately, Bino weighed only ninety-six pounds; the chimp's heart was too small to effectively pump the large amount of blood in Rush's body. Two hours later, Boyd Rush was dead. He never regained consciousness.

Hardy had originally planned to tell the world what he had done two weeks later at a medical conference. But twenty-five people had crammed into Hardy's operating room to witness what they believed was about to be history. One, apparently, leaked the news to the press, which carried the story incorrectly as a human-to-human, not chimp-to-human, heart transplant. The hospital had to issue a correction. Now, news outlets across the nation knew exactly what had happened in that operating room in Jackson, Mississippi.

On February 8, 1964, at the Sixth International Transplantation Conference in New York City's luxurious Waldorf Astoria hotel, James Hardy stood in front of an audience of his peers and described the operation. "It was as if there had been a recent bereavement in my family," recalled Hardy. "There was not a single hand of applause thereafter. It was a dismal day."

Also in attendance at that New York City conference was Dr. Norman Shumway, the surgeon who would later be known as the "father of heart transplantation." Shumway urged restraint. He argued that surgeons needed to improve the preservation of donor hearts and, most important, figure out how to avoid rejection of the donor heart by the recipient's immune system. When answering a question about Hardy's choice to use a chimpanzee heart, Shumway was tactful. "Perhaps the cardiac surgeon should pause," he argued, "while society becomes accustomed to resurrection of the mythological Chimera."

Despite Norman Shumway's warning, between 1964 and 1977 at least four people received hearts from sheep, baboons, or chimps; all died within a few days. The procedure that finally put an end to animal-to-human transplants was performed in a small hospital in Southern California in the 1980s, almost twenty years after the first successful human-to-human heart transplant.

Early in the morning of October 26, 1984, Dr. Leonard Bailey of the Loma Linda Medical Center transplanted a baby baboon's heart into Stephanie Fae Beauclair, a twelve-day-old girl with a severe heart defect. The public knew her as Baby Fae. Bailey had extensive experience transplanting sheep hearts into newborn goats, so he was used to working with small hearts. The operation was a success. At 11:35 a.m. the new heart was beating rapidly in Baby Fae's little body. Although the baboon's heart was the size of a walnut, it did what Stephanie's abnormal heart couldn't do: supply her body with the oxygenated blood needed to keep her alive.

On November 15, 1984, twenty days after the transplant, Stephanie Beauclair was dead. Bailey had hoped that her immune system would be too immature to recognize the baboon's heart as foreign and reject it. It didn't work out that way. Her body sent white blood cells called lymphocytes to the new heart and destroyed it, ending her life. Any hope that surgeons had placed in animal-to-human heart transplants died with Baby Fae.

Animal-rights activists targeted Leonard Bailey. "My family suffered immensely," he recalled. "We had to have police live in our home. Our personal mail was opened by the police department for over a year. I wasn't allowed to appear in public without a bullet-proof vest under my clothing." Although the ethics committee at Loma Linda Medical Center had given Bailey permission to do four more transplants using baboon hearts—and seven baboons were currently housed on site—he never performed another transplant.

The Baby Fae experiment became a cultural icon. Ten years later, in a 1993 episode of *The Simpsons* titled "I Love Lisa," the school cafeteria serves beef hearts in honor of Valentine's Day. Bart puts one under his shirt and declares, "My baboon heart! I'm rejecting it," before throwing it back on the table. And in Paul Simon's *Graceland* album, the song "The Boy in the Bubble" contains the lyric "Medicine is magical and magical is art / Thinking of the boy in the bubble / And the baby with the baboon heart."

———

When James Hardy stood at the podium of the Sixth International Transplantation Conference in New York City in 1964, successful heart transplantation centered on three challenges, two of which had been solved. One has never been solved completely.

First, surgeons had to maintain the health of the donor heart for the hour or so between removal and transplantation. In 1952, researchers at Hahnemann Hospital in Philadelphia found that if they dramatically lowered the temperature of a dog before killing it, the dog's heart didn't suffer any damage. As a result, cooling the donor and recipient before the transplant became standard procedure.

Second, surgeons had to find the best way to attach the new heart to the major blood vessels of the recipient. By 1960, Norman Shumway and Richard Lower at Stanford University had figured that out.

Third, surgeons had to find a way to prevent the recipient from rejecting the new heart. Before transplanting hearts into people, surgeons had spent decades transplanting them into experimental animals. Shumway and Lower were the most successful. They were the first to show that transplanted dogs could live a normal life for a year. Unfortunately, every successful dog-heart transplant ended the same way: the dog's immune system rejected the new heart. "If the immunological mechanisms of the host were prevented from destroying the graft," said Shumway in 1960, "in all likelihood [the heart] would continue to function adequately for the normal life span of the animal." Scientists and surgeons had to find a better way to prevent immune rejection. Although the first successful heart transplant would be achieved just a few years later, it would be twenty years before doctors came close to truly solving this problem.

———

Shumway's observations weren't new. Surgeons had been trying to conquer the problem of transplant rejection for about four hundred years. In the sixteenth century, an Italian surgeon named Gaspare Tagliacozzi found that he could transplant skin from one body site to another without difficulty. The new skin graft would function perfectly. However, if he took skin from one person and transplanted it onto another, the new skin would turn gray, whither, and die.

In the 1930s, Leo Loeb, working with rodents, found that he could successfully transplant skin from one animal to another, as long as the animals were genetically identical. But if they were genetically dissimilar, the animals would reject the transplant—and the greater the dissimilarity, the more vigorous the rejection. Loeb reasoned that transplants in humans could be successful if the donor organ came from an identical twin. As Loeb had predicted, the first successful kidney transplant, which was performed on December 23, 1954, occurred when Ronald Herrick donated a kidney to his twin brother, Richard. Most people, however, don't have an identical twin. Therefore, the only way to prevent the body from rejecting the transplanted organ is with drugs that suppress the immune system.

In 1955, the first immune-suppressive drug, prednisone, became commercially available. In 1963, the second, azathioprine, also became available (and won a Nobel Prize for its inventors). Both drugs lessened but didn't eliminate the possibility of rejection in people with heart transplants. Not surprisingly, both drugs also weakened the body's ability to fight infections. As a consequence, heart-transplant surgeons are always walking a narrow line between preventing fatal rejections and causing fatal infections. No story shows how difficult this high-wire act was than the first human-to-human heart transplant. It was a procedure that for years afterward made the surgeon who did it the most famous doctor in the world—and, later, one of the most vilified.

———

Christiaan Barnard was the son of a poor Dutch Reformed Church minister in Beaufort West, part of the vast open spaces of Karoo, the desert heartland of South Africa. It was Barnard's mother, however, who instilled in Christiaan and his brothers the belief that they could do anything if they just tried hard enough. As a boy, Christiaan set a record for the mile run in his bare feet, won a school tennis championship with cardboard covering the holes in his sneakers, and finished at the top of his class while studying by rural firelight. After graduating from the University of Cape Town in 1945, Barnard won a scholarship to travel to the University of Minnesota, where he became fascinated with heart transplants. (Coincidentally, Norman Shumway, who was the same age as Barnard, trained at the University of Minnesota at the same time.) When Barnard returned to South Africa, he was appointed head of the department of experimental surgery at

the Groote Schuur Hospital in Cape Town, eventually estab-
lishing the first intensive-care unit in South Africa. In 1967,
Barnard spent three months at the Medical College of Virginia to learn more about organ transplantation. When he
returned home, he became the first person in South Africa to
do a kidney transplant.

Barnard was a man of many contradictions. During
apartheid, he ignored the restrictive racial laws in his country, including allowing mixed-race nurses in the operating
room to treat white patients. Nonetheless, he was often bel-
ligerent to his staff. A colleague described him as "egocen-
tric, hard-working, clever, ambitious, brash, and somewhat
arrogant." "I have a tremendous ego," Barnard once told an
interviewer, "and I must feed it, or I become miserable and
unhappy."

In December 1967, Christiaan Barnard would get his
wish—his ego would be fully fed. And it would destroy him.

———

Louis Washkansky was a fifty-four-year-old Jewish grocer
born in Lithuania. When he was twenty-eight years old,
Washkansky, a heavy smoker with diabetes, suffered his
first heart attack. When he was thirty-eight, he suffered his
second; at forty-three, his third. Each left his heart severely
damaged: a grotesque, floppy, quivering balloon. His failing
heart caused massive amounts of fluid to collect in his legs,
which periodically had to be drained. The slightest exertion
left him breathless. "I know as soon as I close my eyes,"
said his wife, Ann, "that he will take one real long breath
and never breathe anymore." Two-thirds of Washkansky's
left ventricle, which is responsible for pumping oxygen-rich
blood to the body, was essentially dead; also, the two arteries that supplied blood to his heart were largely obliterated.

Barry Kaplan, Washkansky's doctor, said it was the largest heart he'd ever seen.

On September 14, 1967, Louis Washkansky was admitted to the Groote Schuur Hospital in Cape Town, breathless and dying. One month later, he suffered kidney and liver failure. The nurse taking care of him said, "He couldn't breathe. He couldn't turn around without puffing. He couldn't do a thing for himself. He was blue. His body was bloated. His legs were draining fluid. He was a very sick man—a very, very sick man indeed." When Christiaan Barnard looked at the X-ray and saw the size of Washkansky's heart, he couldn't believe that he was still alive.

Barnard had his transplant candidate. Now, he had to find a donor.

Twenty-five-year-old Denise Ann Duvall worked at a bank and lived with her parents, Edward and Myrtle. On December 2, 1967, they, along with Denise's brother, Keith, all got into Denise's new car to visit friends. First, they stopped at the Wrensch Town Bakery. "My wife wanted a cake for our friends," recalled Edward, "so we stopped opposite that bakery at Salt River." At 3:35 p.m. Denise and her mother walked out of the shop, cake in hand. Unfortunately, a large truck parked in front of the store blocked their view of the road. When they stepped out from behind the truck, they didn't see the white car driven by Frederick Prins, who had been drinking, speeding toward them. "We heard a thud and a bang and a screech of tires," said Edward. Keith turned around. "Dad!" he shouted. "It's Mom and Denise!" The two women were knocked across the road. Fifty-two-year-old Myrtle was dead on impact. Denise suffered massive head injuries, with blood pouring out of her nose, ears, and mouth. But she was breathing, and her heart was still beating, when the doctors arrived.

Remarkably, at the same time as the accident, Ann Washkansky and her sister-in-law, Grace, were driving out of the parking lot of Groote Schuur Hospital on their way home. Seeing a crowd gathered on the road in front of the Wrensch Town Bakery, Ann slowed down. "Oh, my God," she said, "there's been an accident. There's a woman on the road." "Two women," said Grace.

Denise was brought immediately to Groote Schuur, where she was found to have several large skull fractures. Her pupils were dilated and unreactive to light, evidence of brain death. An electroencephalogram (EEG) showed no brain activity. At the time, South Africa didn't have laws defining brain death. Every hospital had its own rules. "Our [hospital] lawyers stated that a patient could be considered a donor," said Barnard, "once two doctors, one being qualified for more than five years, declared that an individual is dead. It did not state what criteria the doctors had to use. They left that to the medical profession."

Barnard asked Edward Duvall for permission to remove his daughter's heart. "If you can't save my daughter," said Duvall, "you must try to save this man." Then Barnard spoke with Ann and Louis Washkansky. "There is nothing to think about," said Louis. "I'll take the chance as soon as possible." Barnard remembered the moment: "For a dying man, it is not a difficult decision [to accept a heart transplant] because he knows he is at the end. If a lion chases you to the bank of a river filled with crocodiles, you will leap into the water convinced you have a chance to swim to the other side. But you would never accept such odds if there were no lion."

Louis Washkansky idolized Christiaan Barnard, calling him "the man with the golden hands." Ann, on the other hand, didn't trust him. When she asked Barnard about the risks of the operation, he replied that her husband had an

80 percent chance of survival. Ann wondered how Barnard could possibly have known this, given that the operation had never been done before by him or anyone else.

At 12:50 a.m. on December 3, 1967, Louis Washkansky and Denise Duvall were brought into adjoining operating rooms. "I wanted to turn back, but there was no turning," said Barnard. "Both of them had living hearts that could not continue to beat for much longer. We were approaching the moment when there would be nothing else to do other than cut out both of their hearts, and place one of them—the girl's—within an empty chest of the man who would otherwise never leave the operating room alive."

First, Barnard turned off Denise's mechanical ventilator. Standing next to him was his brother, Marius, another heart surgeon. "So, we waited," said Barnard, "while the heart struggled on—five, ten, fifteen minutes . . . until it finally revealed itself in a straight green line across the screen—death. 'Now?' asked Marius. 'No' I said. 'Let's make sure there is no heartbeat coming back." In his account, Barnard didn't want anyone to accuse him later of killing Denise, something that would soon become a problem for American heart surgeons. What really happened in that operating room that morning, however, was somewhat different. Forty years later, after Christiaan Barnard had passed away, Marius revealed that his brother had injected Denise with a potassium solution designed to stop her heart immediately after removal of life support. Barnard didn't want Denise's heart to be damaged from the lack of oxygen that had been supplied by the breathing machine. So, he forced the issue.

Assisted by Marius, Barnard opened Denise's chest, removed her heart, cooled it to fifty degrees Fahrenheit, and, at 2:20 a.m., opened the chest of Louis Washkansky. Holding Denise's smaller heart in his hands, Barnard said, "For a moment I stared at it, wondering how it would ever work. It seemed so small and insignificant—too tiny to handle all the demands that would be put upon it. . . . And the heart of Washkansky had created a cavity [in his chest] twice the normal size. All alone, in so much space, the little heart looked much too small—and very lonely." Using the technique developed by Norman Shumway and Richard Lower at Stanford, Barnard sutured Denise's heart into Washkansky's chest. Initially, the heart lay motionless. So, Barnard gave Washkansky's new heart an electric shock with a defibrillator. "The heart lay paralyzed, without any sign of life," said Barnard. "We waited it seemed like hours until . . . little by little it began to roll with the lovely rhythm of life." Five hours after the start of the operation, Christiaan Barnard could finally take a breath. "Dit werk," he said to Marius in Afrikaans: "It works." Barnard then gave Washkansky medicines to suppress his immune system, hoping to stave off rejection for as long as possible.

At 5:43 a.m., Washkansky was taken back to his room, and at 9 a.m. he woke up. "You promised me a new heart," he told the nurse. "I assume you kept your promise." Washkansky's recovery bordered on the miraculous. "It was amazing to see how he lost all evidence of heart failure," said Barnard. "The swelling of his legs disappeared; the swelling of his liver disappeared; his lungs became dry; and he was well, mentally well. So, we were very optimistic at the beginning."

Louis Washkansky, the first heart-transplant recipient to regain consciousness, was immediately the world's most famous patient. Cabinet ministers, photographers, journalists, and representatives of every major broadcasting organization flooded his room. One particularly low moment came when a BBC interviewer asked Washkansky by phone what it felt like, as a Jew, to have the heart of a gentile. "Well, I never thought of it that way," he replied. "I don't know . . ." One of the doctors in attendance, Dr. Bossie Bosman, cut the connection to London and shouted angrily at a BBC technician, "How do you feel working for a company that asks stupid questions?"

On December 15, 1967, two weeks after the operation, Washkansky worsened. His lungs filled with fluid and he struggled to breathe. Barnard considered two possibilities: either the fluid in Washkansky's lungs was caused by heart failure from rejection, or it was caused by a bacterial pneumonia. Barnard chose the former, dramatically increasing the quantity of immune-suppressive drugs. Unfortunately, he had made the wrong choice. On December 21, eighteen days after the world's first human-to-human heart transplant, Louis Washkansky was dead. The autopsy showed only minimal signs of rejection and lungs full of bacteria. Suppressing the immune system of a man fighting serious bacterial pneumonia was the worst possible thing Barnard could have done. "There was at least a part of my daughter that was still alive," lamented Edward Duvall, "but now she is completely dead."

Heart surgeons across the globe were surprised that Christiaan Barnard, an unknown surgeon from a country that had only recently opened an intensive-care unit, was now basking in the fame of being the first surgeon in history to transplant a human heart. Everyone had assumed that it would have been Norman Shumway taking the bows. "Shumway did everything by the book," said James Hardy, "only to have history stolen from him."

━━

Barnard became an international celebrity. Within two weeks of Washkansky's operation, the surgeon was on the cover of *Time*, *Life*, and *Newsweek* magazines. But he was dissatisfied with the outcome of the transplant. He wanted to prove that he could do better than prolong someone's life for only eighteen days. His next chance came one month later.

The head of the cardiac-care unit at Groote Schuur Hospital was Dr. Velva Schrire. When Barnard informed him that Denise Duvall's heart was available for transplant, Schrire asked one question: "Is she colored?" Barnard's transplant was in 1967. Apartheid was the law in South Africa until 1990. White South Africans, most of whom, like Christiaan Barnard, were of Dutch descent, were known as Afrikaners. South Africans who were Black or mixed race, who composed the majority of the country, were called "coloreds." Apartheid separated Afrikaners and "coloreds" in housing, employment, and public services. Further, the Group Areas Act prohibited Black people from sitting next to white people on a park bench. In the medical realm, apartheid demanded separate wards, separate ambulances, and separate transfusions. An Afrikaner would never receive a blood transfusion from a Black or mixed-race person.

As it turned out, a "colored" man had died two weeks before Denise Duvall. Schrire refused to allow Barnard to use that man to provide a heart for Louis Washkansky. The reason, however, was driven by Schrire's concern about the world's perception of apartheid, not by apartheid laws. "Professor Schrire and I decided that the first patient and donor for a heart transplantation should be white South Africans," said Barnard, "not because of any problems with South African authorities but because we were afraid that if either the patient or donors were Black we would be criticized for having experimented on Black patients."

This fear, however, didn't apply to Barnard's second transplant. Philip Blaiberg was a fifty-nine-year-old retired dentist with a failing heart. "He was gasping for breath," said Blaiberg's sister-in-law. "He was under an oxygen tent all the time. He had no chance of ever living, you know, except day-to-day. If something serious had not been done to him, he would never have lived longer than a week or so. The transplant came just in the nick of time." The donor heart came from a twenty-four-year-old South African of mixed race named Clive Haupt, who had died on the beach following a massive bleed in his brain due to a ruptured aneurysm. Previously, Haupt had been in perfect health.

On Tuesday morning, January 2, 1968, the pathologist at Groote Schuur Hospital signed the death certificate of Clive Haupt. Haupt's mother gave permission for the transplant, echoing the sentiments of Edward Duvall. "If you can save the life of another person," she said, "take my son's heart." Of interest, neither Philip Blaiberg nor Louis Washkansky were of Dutch descent. Both were Jewish, and neither shared the racial prejudices embraced by white Afrikaners.

Under apartheid laws, Clive Haupt would never have been allowed to enter the white wards in a South African hospital. But his heart could, an irony not lost on the press. "There is no provision under the Group Areas Act for black hearts to beat in white neighborhoods," wrote the *Guardian*, a British daily. "There can be no doubt that Mr. Haupt is committing a posthumous offense." Years later, Christiaan Barnard transplanted the heart of a white woman into a Black man.

Philip Blaiberg lived for nineteen months. He kept his old heart in a jar, proudly showing it to anyone who asked. Later, he wrote a book, titled *Looking at My Heart*.

The first heart transplant of a child was also the first human-to-human heart transplant in the United States. To render the issue of brain death moot, the surgeon who performed the transplant—Dr. Adrian Kantrowitz of Maimonides Medical Center in Brooklyn, New York—used a donor who didn't have a brain (anencephaly). About one of every ten thousand babies born in America is anencephalic. Most of these babies are born dead. Some survive a few days or a week, but no longer. The absence of a brain is a fatal condition.

Jamie Scudero was born on November 18, 1967, with a heart defect that prevented blood from traveling to her lungs to get oxygen, which explained why she was blue at birth. On December 6, 1967, three days after Christiaan Barnard transplanted the heart of Denise Duvall into the chest of Louis Washkansky, an anencephalic baby named Ralph Senz was flown into Brooklyn from Portland, Oregon, to provide a heart for Jamie. Kantrowitz had finished the operation by 5:20 a.m. "All these years later I can still see it clearly," he recalled. "It was an unforgettable sight. We had taken the old heart and we needed to move damn fast to fill that huge hole with a new heart. No wonder they were frightened of us. No wonder they thought we were out of our minds." Because Jamie was only three weeks old, Kantrowitz had hoped her immune system would be too immature to reject the new heart. He imagined that he had extended her life for months, if not years.

Seven hours later, however, Jamie Scudero was dead. Kantrowitz didn't claim success. "I think it should be clear to you," he told reporters, "and you should convey it clearly to your readers and your listeners and your viewers that . . . this procedure was an unequivocal failure."

Despite the early deaths, 1968 became the "year of the heart transplant"—more than one hundred were performed across the globe.

At this point in the story, the decision about whether or not to receive a heart transplant was fairly easy. All of the early transplant recipients wouldn't have lived for more than a few weeks had they not had a transplant. They weren't gambling with their lives, because their lives were at an end.

———

One of the greatest research alliances in modern medicine occurred when Norman Shumway and Richard Lower joined forces at Stanford University in Palo Alto, California. Both were Michigan natives. Lower attended Cornell Medical School, moving to Palo Alto in 1958. Shumway, after receiving his medical degree from Vanderbilt University, entered a surgical internship at the University of Minnesota, followed by two years in the US Air Force during the Korean War. He then returned to Minnesota to finish his training. It was there that he first met Christiaan Barnard. They were never friends. "Shumway was a naturally gifted surgeon," recalled one biographer,

> always calm, always able to find humor even in the darkest most challenging cases. Contrast this with Barnard, a grating, driven and, in many ways, tragic figure. Barnard treated many of the people around him poorly, blaming them for errors and bad outcomes. Those who worked with him recognized his brilliance and persistence, but found him annoying, demeaning, and a bit of a phony. Also, unlike Shumway, [Barnard] was not a natural surgeon—he was clumsy and had a tendency to become agitated and nervous when problems arose in the operating room. And he suffered from constant pain and trembling of his hands from rheumatoid arthritis.

In the late 1950s and early 1960s, Norman Shumway and Richard Lower developed a technique that became the gold standard for how to transplant hearts. They were the first to show that with the right amount of immune suppression, dogs could live for more than a year. And they were the first to master the art of lowering the temperature in donor and recipient animals. By the mid-1960s, Shumway and Lower had successfully transplanted more than 300 dogs, 250 more than Christiaan Barnard. And where Shumway and Lower's dogs lived for more than a year, Barnard's were usually dead in ten days.

On January 6, 1968, five weeks after Barnard performed the first and third human-to-human heart transplants, Norman Shumway performed the fourth. Michael Kasparek, a fifty-four-year-old steelworker whose heart had been weakened by a virus, was the recipient. The donor was a forty-three-year-old woman named Virginia White who, like Clive Haupt, had suffered a massive brain hemorrhage. The operation lasted five hours. Michael Kasparek was dead fifteen days later. But Kasparek didn't die from overwhelming infection or rejection; he died from a bleeding stomach ulcer. Forty years after the events in the late 1960s, Norman Shumway was circumspect about not being the first person to do a human-to-human heart transplant. "What is it they always say about being the first?" recalled Shumway. "We all know the first guy to get to the North Pole or the South Pole—or to step across the moon. It's just that second or third guy's name that is a little more elusive. I understand the whole drama of being the first."

In the end, Norman Shumway would be the tortoise to Christiaan Barnard's hare.

Barnard wasn't the first surgeon to do a human-to-human heart transplant because he was the most skilled or the most experienced. He was the first to do it because he lived in a country where the definition of brain death was whatever two doctors said it was. In the United States, brain death wasn't defined and accepted until late 1968, and it wasn't established by law until 1981. Before that, the definition of death was what Norman Shumway called "the boy scout's definition"—you were dead when your heart stopped beating and not before. District attorneys threatened to arrest surgeons who took organs from brain-dead patients whose hearts were still beating.

Months before brain death was finally defined by a group of ethicists at Harvard University, Richard Lower, now at the Medical College of Virginia, was named in a lawsuit that could have ended heart transplants in the United States for at least a decade.

On May 24, 1968, Bruce Tucker, a fifty-three-year-old African American who worked in an egg-packing plant in Richmond, Virginia, fell and hit his head on concrete. He had been drinking that night with a friend when he stumbled. After he arrived at the Medical College of Virginia, an EEG showed no brain activity. A note on his chart read: "His prognosis is grim, his chance of recovery is nil, and his death is imminent."

Joseph Klett Jr. was forty-eight years old and needed a heart transplant. Lower enlisted the help of the Richmond Police Department to find Bruce Tucker's next of kin in order to obtain permission for the transplant. Unfortunately, the police were unable to locate a single family member. Legally, any unclaimed body could be used for medical purposes. So, on May 25, at 2:30 p.m., Tucker was taken off his mechanical ventilator, even though his heart was still beating.

Lower than removed his heart and transplanted into Joseph Klett Jr., who died one week later from massive rejection. Given that one of Tucker's sisters and two of his brothers lived in Richmond—and that one of the brothers worked in a shoe-repair shop only fifteen blocks from the hospital—it's hard to understand why the police hadn't been able to find any of his relatives.

When Tucker's brother Grover came to the funeral home to see his brother's body, a boy who worked at the home said, "Did you know the doctors have taken your brother's heart?" "And, you know, it's a shocking thing," recalled Grover, "taking this heart without permission."

In 1972, four years after Tucker's death, his family sued the Medical College of Virginia and Richard Lower for $100,000. Douglas Wilder, who would later become Virginia's first African American governor, represented them. Wilder argued that Bruce's ventilator had been unplugged not to end his suffering but because he was "unfortunate enough to come into the hospital at a time when a heart was needed." Joseph Fletcher, an ethicist, testified at the trial. "When [brain] function is lost," said Fletcher, "nothing remains but biological phenomena at best. The patient is gone even if his body remains and even if some of its vital functions continue. He may be, technically, 'alive,' but he is no longer human."

The seven-member jury deliberated for a little more than an hour before returning a verdict of not guilty. "I think this [verdict] is important," said Lower. "Had the court decided for some reason that we had to stick to the original definition, then I think virtually every surgeon would have been reluctant to remove a beating heart for a transplant. Patients would have died, and progress would have stopped. This will permit transplantation to continue." Brain death now had a legal precedent.

Two years later, the concept of brain death would have one more day in court. In 1973, Norman Shumway removed a donor heart from a shooting victim. When the attacker came to trial, the defense attorney argued that Shumway was the killer because it was Shumway, not the shooter, who had disconnected the breathing machine from the brain-dead victim. Shumway was absolved of the murder and the shooter was convicted of manslaughter.

———

In January 1968, one month after Christiaan Barnard had performed the world's first heart transplant, Henry Beecher, an anesthesiologist at Harvard University, approached the dean of his medical school. Beecher had recently written an exposé on what he believed was the unethical exploitation of human subjects for medical research. The dean appointed him to chair a committee to "examine the definition of irreversible coma." The committee, which published its findings on July 27, 1968, decided that brain death rested on three pillars: (1) coma; (2) the inability to breathe without mechanical ventilation; and (3) the absence of certain reflexes, such as pupils constricting to light, gagging when a tongue depressor is placed in the back of the throat, blinking when a cotton swab touches the cornea, or grimacing when exposed to painful stimuli. No longer did patients have to meet what Norman Shumway had called the "boy scout's definition" of death. By defining brain death, US surgeons could now remove beating hearts without fear of being sued for wrongful death.

The definition of brain death also had the support of the Catholic Church. Pope Pius XII, in an address titled "The Prolongation of Life," stated that, in a deeply unconscious person whose vital functions were sustained by artificial means, "the soul may already have left the body." The pope also believed that death could only be certified by doctors and was not "within the competence of the Church."

With the definition of brain death now firmly in place, heart-transplant mania spread across the globe. In 1968, hospitals rushed to set up heart-transplant units. Between 1968 and 1970, fifty-two centers in seventeen countries performed 166 transplants. Heart-transplant teams were assembled in Argentina, Australia, Brazil, Canada, Chile, Czechoslovakia, England, France, Germany, India, Poland, South Africa, Spain, Switzerland, Turkey, the United States, and Venezuela. Each of these centers could now claim that they, too, were on the cutting edge of modern medicine. The outcomes, however, were far from reassuring. As a consequence, the number of transplants and transplant centers decreased every year:

- In 1968, 102 transplants were performed. Half of these patients were dead within a month and only 10 percent were alive two years later. Survival times were being measured in hours and days, not years and decades. What the media had hailed as the miracle of Cape Town was darkening.

- In 1969, forty-eight transplants were performed. About one-quarter of the patients were alive one month later and only 6 percent were alive two years later.

- In 1970, only sixteen transplants were performed. About 10 percent were alive after one month and only 3 percent were alive two years later. The number of transplant centers had decreased from fifty-two to ten. In December 1970, the American Heart Association found that of the 166 transplants that had been performed up to that point, only twenty-three patients were still alive.

- In 1971, only seventeen transplants were performed. A medical bulletin asked, "What Ever Happened to Heart Transplants?" On September 17, 1971, the cover of *Life* magazine carried the headline, "The Tragic Record of Heart Transplants." Six transplant patients were shown on the cover. All had died within eight months of having been photographed.

At this point in the story—despite the large number of transplants and transplant centers—the decision about whether to receive a heart transplant hadn't changed very much. For patients with a life expectancy of only a few days or weeks, it was worth taking the risk. If patients could reasonably be expected to live for a couple of years, however, then a heart transplant would likely only shorten their lives. Because mortality statistics were widely available, people could make an informed choice.

People requiring heart transplants in the late 1960s and early 1970s didn't have long lives ahead of them, so it's hard to label their poor survival rates as tragedies. And much was learned. Most important, surgeons now understood that the Holy Grail of transplantation was finding a way to prevent rejection, the most common cause of death following transplant. Until better ways to diagnose and treat rejection became available, the mania to transplant hearts had to subside—which is what happened. There were also some early successes that provided hope. For example, following their heart transplants, Dorothy Fischer lived for thirteen years and Dirk van Zyl survived for twenty-three.

No one worked harder to solve the problem of rejection than Norman Shumway. In 1971, Shumway and the Stanford group published their experience with twenty-six transplants: 42 percent had survived for six months; 37 percent for eighteen months; and 26 percent for two years—statistics dramatically better than those from other institutions. To detect early signs of rejection, Shumway relied on subtle changes in the electrocardiograph (EKG) recording of the patient's heartbeat, as well as the presence of heart-cell enzymes in the bloodstream—evidence of heart damage. This allowed him to find the right mix and timing of immune-suppressive drugs. But Shumway's record of success improved dramatically after he took on a young surgeon from Northern Ireland named Philip Caves.

Caves, much to the surprise of Shumway, chose to spend most of his first few months in Northern California in the Stanford library rather than in the operating room. Caves had an idea. He proposed that the best way to determine whether a new heart was in the midst of rejection was to biopsy it. A biopsy showing mild infiltration of white blood cells would be early evidence of rejection: before the patient's heart started to fail, or before the EKG changed, or before heart-cell enzymes began to spill into the bloodstream. The trick was finding a way to biopsy the heart without hurting the patient.

In 1973, Caves, working with an instrument maker named Werner Schulz, invented the bioptome: a thin piano wire with tiny pincers at the end. The surgeon could now thread this wire into a vein in the neck, snip off a tiny portion of the lining of the heart, and examine it under the microscope. In collaboration with a pathologist named Margaret Billingham, Caves set up a standard grading system to determine when a patient was in the early throes of rejection. Survival rates soared. By the end of the 1970s, four of the seven teenagers transplanted at Stanford had survived for more than ten years.

The field, however, required one more breakthrough.

In 1969, an employee of the Swiss pharmaceutical company Sandoz was vacationing in the mountains of southern Norway. He knew that his company was interested in finding a new antibiotic and he knew that the best place to find it would be in dirt. Bacteria and fungi in the soil compete for nutrients by killing their competitors. These killing agents, called antibiotics, help them to survive. The Sandoz employee filled a small plastic bag with soil and brought it back to the laboratory, where it was found to contain a fungus that made a substance the company code-named 24-556. Researchers at Sandoz were disappointed to find that 24-556 didn't kill any of the bacteria they tested. Clearly, it wasn't going to be the next great antibiotic. But the drug had one property that surprised them.

In 1977, a Belgian physician named Jean-François Borel found that 24-556 suppressed the immune systems of mice and rabbits. Unlike existing immune-suppressive drugs such as prednisone and azathioprine, which suppressed every aspect of the immune system, 24-556 only suppressed the specific type of immune cells that were causing transplants to be rejected. Sandoz researchers found that rabbits given this drug, now called cyclosporine, had kidney transplants that were virtually indefinite. The same was true for heart transplants in pigs. Cyclosporine, it appeared, had dramatically lessened the rejection problem. By 1983, the drug was being used in heart transplants across the globe. The total number of transplants soared from 182 in 1982 to more than 9,200 by the end of the decade. The modern era of heart transplantation had begun.

————

With improvements in donor acquisition, refinement of surgical techniques, breakthroughs in organ preservation, advances in recognition of early rejection, and the development of better drugs for immune suppression, the life span of those receiving heart transplants has continued to increase. One name, however, has been missing from all these advances. The man who was at the center of what was one of the world's most celebrated medical events. A man who, in the course of a single week, had become an international sensation: Christiaan Barnard.

What happened?

A few days after Barnard walked out of the operating room in the early morning hours of December 3, 1968, the media descended upon him. CBS invited him to appear on a special one-hour edition of *Face the Nation*. Following the show, which was seen by more than twenty million Americans, Barnard traveled to Texas to visit President Lyndon Johnson at his ranch. Thousands waited to get his autograph when he landed at airports. Like a politician, he kissed babies who were handed to him. He was touted for the Nobel Prize. One reporter commented that Barnard had "reduced his status as a great scientist to the level of teenage pop idol." Barnard, who had now hired his own publicity director, was proud of it. "I don't think it's bad being an idol," he said. "This is about the first time in history that a scientist and not a pop star, an actor, an athlete, or a boxer became an idol." Whisked into stardom, Barnard didn't attend the funeral of Louis Washkansky, an offense for which Ann Washkansky never forgave him. Washkansky's funeral was televised throughout the world, except in South Africa.

Barnard continued to embrace his fame. After being offered $100,000 for his autobiography, he asked Italian filmmaker Carlo Ponti to make a film about his life, with Paul Newman, Gregory Peck, or Warren Beatty as possible leads. He had lunch with Sophia Loren, was invited to Dean Martin's Hollywood home and Peter Sellers's yacht, and had a very public affair with film star Gina Lollobrigida. "We celebrated with champagne several times during the night and I left early the next morning," said Barnard. "She drove me back to my hotel in her Jaguar—absolutely naked in her mink coat." He was voted "one of the world's five greatest lovers" by *Paris Match*. He traveled to Monaco to meet Prince Rainier and Princess Grace, later flying to Brazil, England, Germany, Iran, Peru, and Spain, as well as Italy, where he was granted an audience with Pope Paul VI. "I mean here's a little boy who was born in Karoo and had a very humble beginning," said Barnard, "and all of a sudden I'm a celebrity. Everybody wants to talk to me; everyone wants to meet me. I get invitations left, right, and center. It was exciting. It was very exciting. I do admit that . . . sometimes I lost my balance because I loved the exposure so much. . . . [But] any man who says he doesn't like applause and recognition is either a fool or a liar."

Barnard's exploits took a toll on his twenty-year marriage to Aletta Louw, a nurse he had married in 1948. The couple had two children: Dierdre, born in 1950, and Andre, born in 1951. In May 1969, they divorced. A few months later, in 1970, at the age of forty-eight, Barnard married a nineteen-year-old heiress named Barbara Zoellner, who was the same age as his son. They also had two children, Frederick, born in 1972, and Christiaan Jr., born in 1974. In 1982, the couple divorced. In 1988, he married a model named Karin Setzkorn. Barnard was sixty-six; she was thirty-eight. They also had two children: Armin, born in 1989, and Lara, born in 1997 (when Barnard's first child was forty-seven years old). That marriage also ended in divorce, in 2000, when Barnard was seventy-eight.

In October 1975, five hundred leading cardiologists and heart surgeons were invited to Detroit for the Henry Ford International Symposium on Cardiac Surgery. Barnard wasn't among them. His colleagues couldn't forgive him his arrogance, his rock-star shenanigans, and his refusal to acknowledge the work of others who had made the first heart transplant possible.

In 1983, Barnard left surgery for good, choosing instead to consult for a "rejuvenation center," which paid him hundreds of thousands of dollars. The clinic "revitalized" its clients with aborted lamb fetuses. He also lent his name to a "miraculous anti-aging cream." Barnard was later stripped of his membership in the American College of Surgeons and its cardiology equivalent. In America and most of Europe, Christiaan Barnard had become an embarrassment to the medical profession.

On September 2, 2001, at the age of seventy-nine, Christiaan Barnard, still lamenting the fact that his third wife had left him, died of an asthma attack while vacationing in Cyprus. Alone. "Show me a hero," wrote F. Scott Fitzgerald, "and I'll write you a tragedy."

Today, about 2,300 heart transplants are performed in medical centers in the United States every year. One-year survival rates now exceed 90 percent, and the average length of survival is fifteen years. Heart transplants are as common as bypass surgery. We've come a long way from the time when we celebrated Louis Washkansky's eighteen days of life. Although better immune-suppressive drugs have further improved the outcome of heart-transplant patients, the problem of rejection remains. Patients still suffer severe and occasionally fatal infections due to these drugs. Also, heart-transplant patients must be maintained on immune-suppressive drugs for the rest of their lives. Perhaps worst of all, more than four thousand people with severe heart disease are currently on the waiting list for a donor, about one-third of whom will die while waiting.

One proposal, however, could solve the dual problems of transplant rejection and long waiting lists. With the advent of genetic engineering and cloning technologies, pigs offer hope for a limitless supply of organs. This idea isn't so far-fetched. Think of Dolly, the world's most famous sheep. Born on July 5, 1996, Dolly was cloned from a single adult cell. Although pig hearts might again evoke the fear of Chimeras and the anger of animal-rights activists, it's a technology whose time might have finally arrived.

For people on the heart-transplant waiting list, the decision of whether to be among the first to receive a pig's heart is again one of weighing relative risks. For several reasons, patients could reasonably choose to receive the pig's heart: (1) heart surgeons have been replacing damaged heart valves with valves from pigs, cows, and horses (called xenografts) for decades, so they've accumulated a lot of experience with animal transplants; (2) pig hearts can be genetically engineered to avoid recognition by the immune system, eliminating the need for immune-suppressive medications that increase the risk of infection—something that people who receive human hearts would still require; and (3) one of every three people on the heart-transplant list will die while waiting.

As you'll see in the next chapter, the choice to use animals to save people didn't end with heart transplants.

CHAPTER 2

RYAN WHITE

Blood Transfusions

EVERY THREE SECONDS SOMEONE IS TRANSFUSED WITH A STRANGER'S blood. In the United States alone, sixteen million units of blood are transfused into ten million people every year. Indeed, blood is more precious than oil. In 2020, oil cost $42 a barrel; the same quantity of blood cost $20,000.

The first three blood transfusions, which were performed in the seventeenth century, received little public attention. The fourth—which has been the subject of hundreds of magazine and newspaper articles and several books—ultimately led to a ban on all blood transfusions. The doctor who did it was indicted for murder. His trial, where the real killer was finally exposed, became a tabloid-style scandal.

———

Richard Lower (no relation to the transplant surgeon from the previous chapter) was born into a farming community in Cornwall, a county on England's rugged southwestern tip, in 1631. Lower, however, wasn't interested in farming, and he later went to Oxford on a scholarship. Although his

41

accomplishment is buried in history, Lower was the first person to make blood transfusions possible.

In 1665, in an experiment backed by the Royal Society in London, Lower removed large quantities of blood from a dog, causing it to go into shock. Then, using a series of tubes, he connected the artery from another dog to a vein of the dog he had just exsanguinated. The spurting artery from the donor dog filled the passive vein of the dying dog, saving its life. Lower had solved one fundamental problem of blood transfusions: clotting. Once exposed to air, blood quickly clots, making it impossible to transfuse. Centuries later, researchers discovered a simpler way to do this. But for the moment, Lower's technique allowed for the first blood transfusions into people.

On June 15, 1667, a twenty-seven-year-old physician named Jean-Baptiste Denis, using the technique developed by Richard Lower, performed the first ever human blood transfusion. Denis had received a bachelor's degree in theology and a doctorate in mathematics before studying medicine in Montpellier, in southern France. Later, after moving to Paris, he became the court doctor to King Louis XIV. Denis also dabbled in medical research.

To perform his transfusion, Denis chose a fifteen-year-old boy who suffered from severe, debilitating fever. First, Denis removed three ounces of the boy's blood. Then, he inserted a thin tube into an artery in a lamb's neck, slipped the other end into the boy's vein, and replaced the three ounces of the boy's blood with the lamb's blood. Denis's reasoning was largely biblical. Blood determined spirit: lambs were calm and weak; stags were courageous and strong. Lamb's blood, therefore, should calm the boy's fever.

Oddly, it worked. At first, the boy felt an intense heat travel up his arm. Five hours later, he had a "clear and smiling countenance." Although he suffered a mild nosebleed, for the first time in two months, he ate and drank well. Denis later hired the boy as his valet, a constant reminder of his breakthrough experiment.

The world's second blood transfusion was performed on a drunken, middle-aged butcher who had been paid for his participation. After receiving blood from a lamb, the butcher jumped off the table, butchered the lamb, threw it over his shoulder, and ran off to a local bar to get drunk, much to the dismay of Denis, who had wanted to monitor his symptoms.

In July 1667, Denis published the results of these two transfusions. His star in the medical community was clearly rising. His third patient, Baron Bond, suffered from a diseased liver and spleen. Forty-five minutes after the transfusion, this time from a calf instead of a lamb, Bond sat up and drank some broth, later gradually improving.

Then came the transfusion to end all blood transfusions, at least for the next two hundred years.

—

Antoine Mauroy was thirty-four years old when he became the fourth person to receive a blood transfusion. For years, he had suffered bouts of insanity lasting ten months or longer. During these episodes, Mauroy would beat his wife, Perrine, then run naked through the streets, setting fires along the way. Sometimes he would stop, flail his arms, and let out a murderous howl—much to the delight of local children, who gleefully paraded behind him. To treat these outbursts, physicians had prescribed baths with different combinations of herbs, chemicals, and other "active" ingredients. Special potions had been strapped to his forehead. All without relief.

One day, a nobleman, finding Mauroy wandering naked through the streets of Paris, brought him to the home of Jean-Baptiste Denis. On December 19, 1667, at 6 p.m., Denis opened a vein above Mauroy's elbow with a silver tube, removed ten ounces of blood, and replaced them with blood spurting from an artery on the inner thigh of a calf. Mauroy dozed in a chair for a few hours, then asked for some food. He spent the night sleeping and occasionally whistling, clearly more subdued than before the transfusion. Two days later, Denis repeated the procedure, this time causing a severe reaction. At first, like the fifteen-year-old boy who had been transfused before him, Mauroy complained of an intense heat traveling up his arm. His pulse quickened and his temperature rose; drenched in sweat, he vomited up the bacon and fat he had just eaten. His urine was pitch black, "as if it had been mixed with the soot of chimneys." Mauroy then slept for ten hours, awaking refreshed. Still, he continued to produce black urine. On Friday, December 23—two days after the transfusion—he bled so badly from his nose that a priest was summoned to take his confession. Despite this setback, he recovered, calmer and saner than he had been in eight years. Reunited with his wife, Mauroy was said to have gone from "full-moon lunatic to calm man." For all appearances, the two blood transfusions had worked.

Over the objections of Denis, who had wanted to continue to observe him, Perrine took her husband home. Soon, however, Mauroy regressed, again beating Perrine and returning to his old habits of drink and debauchery. Perrine brought Mauroy back to Denis, showed him the bruises on her face, and insisted on a third transfusion. At this point, the details of the story become fuzzy. One thing, however, was clear: the day after Perrine asked for the third transfusion, Antoine Mauroy was dead.

On April 17, 1668, Jean-Baptiste Denis was tried in a Parisian court for the murder of Antoine Mauroy. The prose-

cutor spoke first: "The case against Monsieur Denis is that in
the last days of January this year, he and his accomplice did
unlawfully kill a patient, one Monsieur Antoine Mauroy. The
accusation is that, despite protestations from learned gentlemen within the Faculté de Médicine, he performed a series of
unnatural operations, transfusing blood from calves into the
victim's veins. He did this not once, but three times. The day
after the third transfusion being the day the patient died. The
inference is clear. Denis killed Mauroy."

Denis argued that Mauroy's third transfusion—the one
that had supposedly occurred the day before his death—had
never actually happened. He said that although he had begun
the procedure, he had immediately stopped when Mauroy
had a seizure. Not a single drop, pleaded Denis, had been
transfused.

But Denis had an even better defense. One witness
claimed that they had seen someone pay Perrine to kill her
husband. Another witness claimed that he had seen Perrine
feed her husband a broth that, when spilled on the floor, had
caused their cat to keel over dead, presumably from arsenic
poisoning. Also, the week before Mauroy died, several people
had heard him say that his wife was trying to kill him. When
the court ruled that Perrine had poisoned her husband, Jean
Baptiste Denis walked free. Perrine was executed.

It is, in retrospect, remarkable that witnesses had seen
Perrine receive a payment from a stranger, throw a ladle on
the floor, and kill the family cat. One fact, however, was unde-
niable: Perrine Mauroy had recently come into some money.
Despite her extreme poverty, she had paid for the calf that
was to be used for the transfusion, as well as for her husband's coffin, grave, and gravedigger. Who would pay Perrine
to kill her husband? Many in the French medical community
were violently opposed to blood transfusions, which they

viewed as unnatural. Apparently, this group was so desperate to bring an end to the practice that it was willing to frame Denis by paying for a hit job on Mauroy. Or so it seemed.

Antoine Mauroy's death put an end to legal blood transfusions on the European continent. In 1667—the same year that Denis performed his first transfusion—Pope Innocent XI signed an order banning the procedure. Two years later, the French parlements banned them. Eleven years after that, the Parliament of England did the same. These bans, however, didn't stop all blood transfusions. They just sent them underground. Indeed, animal-to-human transfusions were performed well into the early twentieth century.

Although the events surrounding the death of Antoine Mauroy remain somewhat cloudy, one thing was clear: Mauroy improved after his transfusions. How was this possible? In all likelihood, Mauroy's psychosis was the result of a syphilis infection of his brain, which could account for all of his symptoms. One of the side effects of Mauroy's transfusions was fever. It is now well established that the bacterium that causes syphilis (*Treponema pallidum*) is exquisitely sensitive to higher temperatures. Indeed, in 1927, Julius Wagner-Jauregg received a Nobel Prize for proving that syphilis could be treated with fever therapy, either by placing patients in a "fever cabinet" or by injecting them with malaria parasites before treating them with lifesaving quinine. Perhaps most surprising, malaria therapy for syphilis existed well into the 1950s.

Unfortunately, nothing was learned from the failures of these early blood transfusions. Not until a Nobel Prize–winning discovery was made.

———

Early animal-to-human blood transfusions were often associated with fever, chills, lower back pain, darkened urine, and a warm, burning sensation at the site of injection. Today, these symptoms would be called transfusion reactions. What causes them?

Karl Landsteiner was a young researcher working at the Institute for Pathological Anatomy in Vienna, Austria, when he became curious about a phenomenon he couldn't explain. If blood is allowed to clot, the red blood cells and white blood cells sink to the bottom of the tube. The straw-colored portion that floats to the top is called serum. When Landsteiner took serum from one person and added it to the red blood cells of another, he found that sometimes the red blood cells would clump and burst; other times, they wouldn't.

Landsteiner then did an experiment that was surprisingly simple, remarkably easy to perform, took very little time, and won a Nobel Prize. In 1901, he took the serum and red blood cells from twenty-two colleagues and fellow lab workers. He then identified three different patterns of clumping. He found that serum from people with what he called type A blood would clump red blood cells from people with type B blood and vice versa. Serum from people with type C blood (which was later called type O) clumped the red cells from those with types A or B. One year later, some of Landsteiner's students identified a fourth group, called AB. Serum from people with type AB blood didn't clump the red blood cells from those with type A, type B, or type O blood. At the time that Landsteiner performed his experiments, scientists had assumed that the clumping of red blood cells occurred because either the red blood cells or the serum had been taken from people who were sick. Landsteiner proved otherwise. Normal red blood cells from healthy people had different proteins on their surface (A, B, or AB); these proteins could be recognized by antibodies from someone else's blood. When they were recognized, the red blood cells could be destroyed, causing transfusion reactions that were potentially fatal.

Landsteiner wasn't finished. In 1919, he left Vienna and traveled to New York City to work at the Rockefeller Institute. While there, he took blood from rhesus monkeys and injected it into rabbits and guinea pigs, which allowed him to identify yet another protein on the surface of red blood cells called Rh (for rhesus monkey). This finding helped explain why some blood transfusions thought to have been with the right type of blood had still caused serious reactions. People with Rh negative blood can't receive blood from someone who is Rh positive (about 85 percent of people are Rh positive). This is especially a problem during pregnancy when mothers who are Rh negative are carrying a baby who is Rh positive. The Rh-negative mother can react against her baby's blood while the baby is still in the womb, with occasionally fatal results. This problem was so severe that until a solution could be found—inoculation of mothers with a product called RhoGAM—couples were prohibited by law to marry if the woman was Rh negative and the man was Rh positive.

In 1907, Reuben Ottenberg, a twenty-five-year-old doctor at New York's Mount Sinai Hospital, became the first person to transfuse blood from one person into another using Landsteiner's typing technique. Ottenberg found that both the patient and donor had type O blood. The era of safer blood transfusions had begun. In 1930, Karl Landsteiner won the Nobel Prize in Physiology or Medicine for "his discovery of human blood groups." His work made modern-day surgery possible.

Blood typing also became the first scientific test used to determine paternity. One particular celebrity scandal brought blood typing to international attention. In 1941, Charlie Chaplin, best known for his loveable, childlike Tramp character, met a promising young actress named Joan Barry. Although Chaplin was still married to his third wife, *Modern Times* actress Paulette Goddard, he and Barry became lovers. Barry later gave birth to a daughter, Carol Ann, and claimed that Chaplin was the father. At trial, Chaplin's lawyers asked for a blood test using Karl Landsteiner's technique. Chaplin was type O; Barry was type A; and Carol Ann was type B. Therefore, Chaplin couldn't possibly have been the father, who had to have been type B. For Charlie Chaplin, that was the good news. The bad news was that, although blood typing was the only test for paternity available at the time, it was inadmissible in court. In April 1945, a jury, by a vote of eleven to one, ruled that Chaplin was Carol Ann's father, even though he wasn't. He was forced to pay child support and legal fees. In 1946, Chaplin appealed the verdict but lost.

Chaplin's case changed paternity laws in America. In 1953, California, New Hampshire, and Oregon became the first states to draft the Uniform Act on Blood Tests to Determine Paternity, which stated that if "the scientific evidence proved that the alleged father is not the father of the child, the question of paternity shall be resolved accordingly." Charlie Chaplin, nonetheless, was forced to continue to pay child support for Carol Ann.

———

By the early 1900s, human-to-human blood transfusions had become more common. George Crile, a Cleveland surgeon, was one of the first to do them. In December 1905,

Crile linked the vein of a twenty-three-year-old woman to an artery in the arm of her husband. Several hours later, the woman was dead.

The problem with Crile wasn't that he was a poor surgeon. Indeed, his vein-to-artery linkages worked perfectly. The problem was that he had completely ignored the work of Karl Landsteiner. In 1909, eight years after Landsteiner had defined blood types and two years after Reuben Ottenberg had used blood typing to safely transfuse a patient in New York City, Crile said that, "contrary to common belief, normal blood of one individual does quite as well as that of another. Kinship apparently is of no special advantage." But Crile often used relatives as donors, lessening the chances of severe transfusion reactions; indeed, two-thirds of Crile's donors were relatives. And when the recipient was a child, the donors were always the parents. Crile's denial of the importance of blood types meant that some patients died from severe transfusion reactions. After the death of a thirty-three-year-old man who had received blood from his brother-in-law, only to die two days later, Crile was finally censured by his colleagues.

Blood typing didn't become routine until after World War I. And it wasn't until 1913 that syringes, needles, stopcock devices, and glass tubes coated with paraffin eliminated the need for direct artery-to-vein transfusions. Now blood could be drawn up into a syringe from one person and injected directly into the vein of another. Still, another problem remained: clotting.

——

It is remarkable—at least in retrospect, after centuries of attaching human veins to lamb arteries, calf arteries, and other human arteries, all to prevent blood from clotting—how easy it was to solve the problem.

In 1914, Dr. Richard Lewisohn, working at New York City's Mount Sinai Hospital, added sodium citrate to blood and found that no matter how long it was exposed to air, the blood remained fluid. And he only needed a 0.2 percent solution of sodium citrate to do it. This single discovery allowed blood to be stored. No longer did patients have to be transfused with the blood of someone sitting right next to them.

The first transfusion center appeared in a Leningrad hospital in 1932. Five years later, Cook County Hospital in Chicago established the second one, coining the term "blood bank." Not only could doctors now store blood, but they could also separate the solid component of blood (red blood cells for carrying oxygen, white blood cells for fighting infections, and platelets for clotting) from the liquid component (plasma)—all just ahead of the war to end all wars.

During World War II, America's support of the Allies extended well beyond supplying arms, aircraft, missiles, and boats. Americans also sent their blood overseas. No one was more important to this effort than a man who later became the subject of an episode of the popular television show M*A*S*H, as well as a cautionary tale from a well-known comedian.

Charles Drew graduated from Amherst College in 1926 but didn't have enough money to go to medical school. So, for the next two years, he worked as a biology teacher and coach at Morgan College (now Morgan State University) in Baltimore. In 1928, with funds in hand, he attended medical school at McGill University in Montreal, graduating second in his class.

In 1938, Drew received a Rockefeller fellowship to study at Columbia University and train at the Presbyterian Hospital in New York City, where he worked on a method for processing and storing plasma. (Plasma is serum that also contains the proteins necessary for clotting.) His research formed the basis of his thesis, titled "Banked Blood." In 1940, Drew became the first director of an international program for the shipment of blood products. Called the Plasma for Britain program, Drew's task was fraught with danger. Although it wasn't hard to separate plasma from blood—even from the seventeen thousand units he had initially received—Drew had to make sure that not a single unit of plasma was contaminated with bacteria; otherwise, his shipments to Britain could be deadly. Plasma is a rich medium in which bacteria can grow.

Drew also had to make clear on the label the type and source of the blood. Many white soldiers didn't want to receive blood from African Americans. This was particularly distasteful to Drew, who was the first African American to receive a doctorate from Columbia. By the time the war ended in 1945, the American Red Cross had collected more than thirteen million units of blood (a unit is about one pint). Every unit was marked by race to distinguish its source. This was life in Jim Crow America.

On April 30, 1950, Charles Drew and three other African American physicians were driving through the night to a gathering in Tuskegee, Alabama. Just before 8 a.m., while driving along Route 49 in North Carolina, the car veered off the highway, hit a bump, and rolled over three times. Drew was thrown out of the car, which then rolled on top of him, causing massive crush injuries. All four men were taken to the Alamance General Hospital in Burlington, North Carolina, which was racially segregated. The emergency department, however, served both Black and white patients. The doctors on call quickly attended to Drew's injuries, but to no avail. Charles Drew was dead. He was only forty-five years old. Although Drew didn't discover the method for separating plasma from blood, he had scaled up the procedure and made it safe, no doubt saving tens of thousands of lives during World War II—a miraculous achievement. That's the story of Charles Drew. Unfortunately, it's not the one that is so often told.

In the 1960s and 1970s, comedian Dick Gregory occasionally informed audiences that Charles Drew, the man at the center of blood banking and blood distribution during World War II, had died in a segregated North Carolina hospital that had refused to provide a lifesaving blood transfusion. Both *Time* magazine and the *New York Times* gave credence to the story. Nothing, however, popularized this apocryphal tale more than the immensely popular television show *M*A*S*H*, about a mobile army surgical hospital during the Korean War. In one episode, a wounded soldier wakes up to find that he has received a blood transfusion. Worried, the soldier asks the doctors whether he has received "white" blood. At night, while he is sleeping, the doctors paint his face with iodine, causing him to believe, idiotically, that "Black" blood has darkened his skin. Then the doctors tell him the incorrect story about Charles Drew and the North Carolina hospital's refusal to transfuse the man who had virtually invented blood transport.

The truth, however, is different. When his car rolled over, Drew suffered a broken neck and a crushed chest, and a vein that carries blood to the heart was severed. The doctors in the emergency department had done everything they could. But, as noted by one of Charles Drew's fellow passengers at the time: "All the blood in the world could not have saved him."

By the early 1940s, blood for transfusion existed in two forms: whole blood, which lasted for about a week in the refrigerator, and plasma, which was far too easily contaminated with bacteria. As the United States entered World War II, a chemist named Edmund Cohn—with the encouragement of the American Red Cross—found a way to better mobilize blood for the battlefield. Working in the basement of Harvard Medical School, Cohn exposed whole blood to different concentrations of alcohol and salt, different temperatures, and different levels of acidity (pH). He found that various fractions of blood precipitated in much the same way that petroleum fractionates to produce gas, oil, and other products. Cohn assigned a number to each fraction. The most important was fraction number 5, which contained albumin, the lifesaving protein in plasma that can restore blood pressure in the face of massive bleeding. Best of all, the albumin precipitated as a dry, white powder that was easily stored and had a much longer shelf life. The first test of Cohn's albumin followed the attack on Pearl Harbor, where it saved the lives of many wounded soldiers.

Other fractions contained various amounts of antibodies (proteins that help fight infections), cholesterol, and clotting factors. Forty years after Cohn described his technique, the fraction of blood that contained clotting proteins would be responsible for one of the worst biological disasters in history: a disaster that almost eliminated an entire group of children with one particular disease.

Clotting wasn't the only problem with blood transfusions that remained unsolved. Sometimes blood contains infectious agents, such as bacteria, viruses, fungi, or parasites. By the late 1930s, measles, malaria, and syphilis had been reported following blood transfusions; many of these cases were fatal. All of these transfusion fatalities, however, paled in comparison to a single outbreak that occurred in the 1940s.

In March 1942, the Office of the Surgeon General noted a growing incidence of jaundice (yellowing of the skin caused by liver disease) among US Army personnel stationed in California, England, Hawaii, Iceland, and Louisiana. All of those jaundiced had recently received a yellow fever vaccine, which, in addition to containing yellow fever vaccine virus, contained human serum as a stabilizing agent. On April 15, 1942, the surgeon general ordered that yellow fever vaccination be discontinued and that all existing lots be recalled and destroyed. Shortly thereafter, manufacturers made a yellow fever vaccine with water instead of serum, but it was too late.

The serum used to stabilize the yellow fever vaccine had been obtained from nurses, medical students, and interns at Johns Hopkins Hospital in Baltimore, several of whom had a history of jaundice and one of whom was actively infected at the time of the donation. By June 1942, fifty thousand US servicemen had been hospitalized with severe liver disease, and 150 had died from what would later be known as hepatitis B. Of the 141 lots of yellow fever vaccine provided to the army, seven were definitely contaminated. Among those who received one of those seven lots, 78 percent became infected. When the dust settled, 330,000 servicemen had been infected and one thousand had died. This was then and remains today one of the worst single-source outbreaks of a fatal infection ever recorded.

Was the yellow fever vaccine tragedy preventable?

In 1964, Baruch Blumberg discovered hepatitis B virus. By 1971—thirty years after the tragedy—a blood test was available to detect it. In 1972, the FDA mandated that all blood be screened for the presence of the virus. So, the army can't be blamed for not screening the blood of donors with a test that wouldn't become widely available for another thirty years. Nonetheless, the choice to use nurses, medical students, and interns as blood donors was a bad one. Medical personnel are more likely to be exposed to people with contagious diseases. Also, one of the medical students was actively jaundiced at the time of donation. No one had asked this student about his current illness on the intake form. This was unforgivable. From the earliest days of blood banking, asking donors whether they were sick was standard practice. The hospitalization and deaths of thousands of US soldiers caused by hepatitis B in 1942 was entirely preventable.

———

Hepatitis B wasn't the only hepatitis virus that could contaminate donated blood. The other—which was initially called non-A, non-B hepatitis, because researchers only knew that it wasn't either one of those—was later called hepatitis C virus. By the late 1970s, hepatitis C virus was found to cause about 90 percent of blood-transfusion-associated hepatitis. In 1984 alone, hepatitis C infected 180,000 people who had received blood transfusions, killing 1,800. The popular entertainer and comedian Danny Kaye was one such victim. It wasn't until 1986 that a blood test was available to detect hepatitis C virus in donated blood.

But there would be another problem beyond syphilis, measles, malaria, hepatitis B, and hepatitis C. In the late 1970s, a different virus entered the US blood supply. This particular virus was so feared, so vilified, and so misunderstood that citizens worried that they could catch it not only from *receiving* donated blood, but from the mere act of *donating* it. As a consequence, blood donations in the United States plummeted. Remarkably, one-third of Americans today still believe that people can catch this virus by donating blood.

In June 1981, five healthy gay men developed an unusual pneumonia caused by a fungus called *Pneumocystis carinii* that had previously been found only in people with cancer. By the end of the year, a hundred more cases had been reported. The syndrome was initially called gay-related immune deficiency (GRID). By May 1982, 355 cases of GRID had been reported and 136 people had died. The disease primarily struck gay men living in New York and San Francisco.

No one knew what caused it. Because the disease primarily affected gay men, the gay lifestyle was put under a microscope. Was there something in sperm ejaculated into the rectum that was suppressing the immune system? Was the recreational drug amyl nitrite (poppers), used during intercourse to enhance sexual pleasure, somehow weakening the immune system? Could this be another sexually transmitted viral or bacterial infection, like syphilis, gonorrhea, or chlamydia? The difference with this particular infection, however, was that it was damaging the immune system. Other sexually transmitted infections didn't do that.

In January 1982, following the death of a sixty-two-year-old man from *Pneumocystis carinii* pneumonia, a physician in Miami phoned Dr. Bruce Evatt at the Centers for Disease Control and Prevention (CDC). The man who had died wasn't gay but had received a blood product. Evatt knew that the blood product had been filtered, which would have eliminated any possible bacterial, fungal, or parasitic contamination. He reasoned that the germ that had caused this man's disease had to have been small enough to pass through the filter—in other words, a virus. Evatt also feared that the virus had now entered the blood supply. Because this man wasn't gay, the acronym GRID was changed to AIDS, acquired immune deficiency syndrome.

By March 1983, more than 1,200 cases of AIDS had been reported, including seventeen possible transfusion cases. By the end of 1983, more than three thousand cases and 1,300 deaths were reported. That same year, a French researcher named Luc Montagnier isolated the agent that was causing AIDS, later called human immunodeficiency virus (HIV). In August 1984, an experimental test to detect HIV was developed. By April 1985, this test was routinely being used by blood banks across America. But the damage had been done. Between 1978 and 1985, twenty-nine thousand Americans who had received tainted blood transfusions had developed AIDS, and most of them would later die from the infection. One group, however, suffered more than any other.

Historically, the canary in the coal mine for tainted blood has always been patients who are the most reliant on blood donations: hemophiliacs. As a consequence, no group of patients was hit harder by the presence of HIV in the blood supply than this one.

Hemophilia is one of the oldest hereditary diseases on record. In the 1920s and 1930s, it received international attention as a disease that affected European royalty. It is caused by the lack of one of two proteins (factor VIII or factor IX) responsible for clotting. As a consequence, blood from hemophiliacs doesn't clot, even when exposed to air. Typically, the disease becomes apparent in children when bleeding doesn't stop following a circumcision, teething, or minor cuts or bruises while crawling. Hemophilia occurs almost exclusively in boys and men; one of every five thousand American males suffers from this disease.

People with hemophilia are treated with a fraction of plasma (called cryoprecipitates) made possible by Edmund Cohn's work at Harvard. The good news about cryoprecipitates is that they are rich in clotting proteins like factors VIII and IX. Indeed, the availability of cryoprecipitates increased the life span of hemophiliacs by twenty years. The bad news was that these fractions were often pooled from many donors, typically about ten. (Transfusions of whole blood, on the other hand, come from one donor.) The use of many donors as a source of cryoprecipitates dramatically increased the risk that one of these donors was infected with HIV.

At the start of the AIDS epidemic, American boys with hemophilia lived as long as those without the disease. By the end of the 1980s, among the ten thousand American males with severe hemophilia, nine thousand were infected with HIV. By 1994, more than 25 percent of the American hemophiliac population had died from AIDS. Most were children and adolescents.

No one case, however, exemplified the fear, ignorance, hostility, loathing, and mistrust generated by this disease more than that of a remarkably brave young boy from Indiana—a boy who became the poster child for the AIDS epidemic.

Ryan White was born on December 6, 1971, at St. Joseph Memorial Hospital in Kokomo, Indiana, to Hubert and Jeanne White. When he was three days old, Ryan was diagnosed with hemophilia when his circumcision failed to stop bleeding. From that point forward, he received weekly transfusions of cryoprecipitates. In December 1984, when he was thirteen, he developed severe pneumonia and was diagnosed with AIDS. He was told that he had six months to live.

After his bout with pneumonia, Ryan was too ill to go to school. By 1985, however, he started to feel better. Jeanne asked the school board whether it would be all right for Ryan to return to his sixth-grade class, but was told that he could not. On June 30, 1985, the Whites filed a formal request to permit their son to attend school. Again they were denied, this time by Western School Corporation superintendent James O. Smith, sparking an appeal process that lasted eight months. Parents of children at Ryan's school, however, weren't interested in waiting for the results of the appeal. Along with fifty teachers, more than one hundred parents signed a petition to bar Ryan from attending school, even though the Indiana state health commissioner, Dr. Woodrow Myers, in addition to several health officials at the CDC, had told the school board that Ryan posed no risk to other students.

But fear ruled.

On November 25, 1985, the Indiana Department of Education determined that Ryan White should be allowed to attend school. On December 17, the school board, by a vote of seven to zero, overturned the ruling. Three months later, the Department of Education again ruled that Ryan could attend school, which he did; after two weeks, however, a different judge granted a restraining order to prohibit his attendance. That same month, the *New England Journal of Medicine* published a study of one hundred people who had lived in close nonsexual contact with people with AIDS and concluded that the risk of catching the disease was nonexistent, even when contact included sharing toothbrushes, clothing, combs, and drinking glasses, sleeping in the same bed, and hugging and kissing. On March 2, 1986, parents of Ryan's classmates held an auction at the school to raise money to continue the fight against Jeanne and Hubert's wish that their son be treated like a normal person, not a pariah. On April 10, circuit court judge Jack R. O'Neill dissolved the restraining order, and the Indiana Court of Appeals refused to hear any further appeals by the parents who wanted to continue to keep Ryan White from attending school.

In April 1986, Ryan White was admitted to the eighth grade. In response, 151 of Western Middle School's 360 students stayed home, and a handful transferred to other schools. Threats against Ryan and his parents continued. Children at school would yell, "We know you're queer." People on Ryan's paper route canceled their subscriptions, believing that HIV could be transmitted by touching a newspaper. When editors and publishers at the *Kokomo Tribune* supported Ryan both editorially and financially, they were ridiculed by the community and threatened with lawsuits.

Although Ryan attended Western Middle School for the entire 1986–1987 year, he was deeply unhappy and had few friends. The school required him to eat with disposable utensils, forced him to use a separate bathroom, and prohibited him from participating in gym class. People refused to shake Ryan's hand or sit next to him in church. His family's car tires were slashed. When a bullet was fired through the Whites' living room window, they moved to Cicero, Indiana, so Ryan could attend the Hamilton Heights High School in Arcadia. On August 31, 1987, Ryan, understandably nervous about how he would be received, was greeted with open arms by school principal Tony Cook, school superintendent Bob Carnal, and a handful of students—all of whom had been educated about HIV and how it was and wasn't transmitted.

Between 1986 and 1987, Ryan White appeared on the CBS *Evening News*, ABC's *Good Morning America*, and NBC's *Today Show* and *Tonight Show*, and he was the subject of feature-length stories in *USA Today* and *People* magazine. Indeed, no one in this country did more to educate the public and the press about AIDS in the late 1980s than this brave, persistent young boy, who continued to project normalcy in the face of a death sentence. Singers John Mellencamp, Elton John, and Michael Jackson, diver Greg Louganis, Surgeon General C. Everett Koop, Indiana basketball coach Bobby Knight, basketball star Kareem Abdul-Jabbar, and actress Elizabeth Taylor all befriended him.

On March 3, 1988, Ryan White spoke before the President's Commission on the HIV Epidemic, describing his different experiences in Kokomo and Cicero to show the importance of education about the disease. That night, he simultaneously appeared on CNN and ABC's *Nightline*, insisting that he was not a hero, just someone who wanted to be treated like a normal kid. In 1989, ABC aired *The Ryan White Story*, starring Lukas Haas. On March 29, 1990, Ryan was admitted to Riley Hospital for Children in Indianapolis with severe pneumonia. He never recovered. On April 11, 1990, almost six years after he had been told that he had six months to live, this courageous young man was laid to rest in front of 1,500 people at the Second Presbyterian Church in Indianapolis. His pallbearers included Elton John, football star Howie Long, and television talk-show host Phil Donahue.

Four months after his death, the US Congress passed the Ryan White Comprehensive AIDS Resources Emergency (CARE) Act to provide support to persons afflicted with this disease. Up to that point, many Americans had seen AIDS as a punishment for bad behavior. Ryan White had helped change that perception. Everyone who was infected with HIV, independent of their sexual orientation, Ryan argued, was an innocent. No one deserved to die from an infection. The CARE Act, which has been reauthorized twice, is the largest program providing services for people living with AIDS in the United States.

In the year following his death, Ryan White's grave site was vandalized four times.

———

Because the response to the epidemic was delayed, many of those who died from AIDS in the 1980s died needlessly. Several factors contributed to this.

First, the reluctance by public health officials to disallow certain high-risk groups—such as gay men with multiple partners and intravenous drug abusers—from donating blood early in the epidemic slowed the response. The reasoning behind this delay was political, not scientific. Members of these groups felt that they were being unfairly stigmatized and shunned.

Second, after World War II, France, the Netherlands, and the United Kingdom, but not the United States, had a solely volunteer blood-donor system. Because volunteer donors have no reason to lie about their health, this made for a cleaner blood supply. In 1978, the FDA required all blood to be labeled as volunteer or paid, and payment for blood eventually died out. Several companies, however, continued to pay for plasma donations, the source of clotting factors for hemophiliacs like Ryan White.

Finally, Ronald Reagan refused to mention the word AIDS until 1987, six years after the medical community had recognized its existence. Had Reagan used his considerable platform to raise public awareness, the problem of a contaminated blood supply could have become a greater national priority.

In the end, however, although a different and earlier response would have lessened the impact of the AIDS epidemic, it wouldn't have prevented the massive contamination of America's blood supply with HIV.

In the wake of the HIV tragedy, several changes have been made to the handling of blood and plasma. Today, the requirement for heat, solvent, and detergent treatment of blood has dramatically reduced the likelihood of contamination with certain viruses. Indeed, no cases of hepatitis B virus, hepatitis C virus, or HIV have been associated with plasma products since 1985.

Transfusion of whole blood, however, is a different story. Although the routine testing of whole blood for the presence of bacteria such as syphilis and viruses such as hepatitis B, hepatitis C, HIV, West Nile, and Zika, as well as two unusual parasites, has decreased the possibility of transmitting these infections, it hasn't eliminated it. For example, before routine blood testing of whole blood became routine

- the risk of getting HIV was one per 2,600 transfused patients—today it's less than one per million (indeed, only one case of HIV has been found in more than thirteen million units of transfused whole blood);

- the risk of getting hepatitis C virus was one per one hundred thousand transfusions—today it's less than one per million; and

- the risk of getting hepatitis B virus was one in sixty-three thousand transfusions—today it's eight per million.

In no case, however, is the risk from a transfusion of whole blood zero. Many other potential agents might be present in blood for which routine testing isn't performed.

Epstein-Barr virus (the cause of mononucleosis), cytomega-lovirus (which causes birth defects), parvovirus B19 (which causes a rash, fever, and anemia), Ebola virus, dengue virus, chikungunya virus, SARS, MERS, SARS-CoV-2, and prions (which cause mad cow disease) are not subject to routine testing. And bacteria, such as salmonella among others, are still the most common cause of death following blood trans-fusions. In short, the receipt of a stranger's blood will always involve some measure of risk.

One would imagine, however, that it shouldn't be too difficult to make red blood cells in a laboratory. This artificial blood could do what red blood cells normally do, carry oxygen to different parts of the body. How hard could that be? The answer, apparently, is very hard.

Artificial blood, which has been described in scientific and medical journals for more than a decade, is still too new. We are going to have to wait until the technology is further along before we can feel comfortable using it. Choices to use new technologies such as artificial blood invariably boil down to determining when we know *enough* to say that a therapy's potential benefits outweigh its known or theoretical risks. We can't wait until we know *everything*, because we never know everything.

Studies of more and more people reduce uncertainty, but they never eliminate it. How many people would need to be transfused with artificial blood safely before we feel that we have critically reduced the level of uncertainty? Ten? Fifty? One hundred? One thousand? As more and more people receive artificial blood, uncertainty will be reduced. But it will never be eliminated.

The same calculation could be made for any new vac-cine, such as the COVID-19 vaccines. When do we get past the point where we have eliminated a critical amount of uncertainty? Phase 1 studies of a new vaccine involve twenty to one hundred people; phase 2 studies, several hundred; phase 3 studies, tens of thousands. Each of these steps provides a greater level of certainty. But ten thousand people isn't ten million. When do we cross the line into relative certainty? Do we ever cross it?

CHAPTER 3

HANNAH GREENER

Anesthesia

THE STORIES OF HEART TRANSPLANTS AND BLOOD TRANSFUSIONS dramatize the risks of new technologies, some of which are avoidable, others not. For example, unavoidable risks include human error (such as the contamination of the yellow fever vaccine with serum containing hepatitis B virus), lack of knowledge (blood group typing wasn't available until 1907), or the sudden and unpredictable appearance of unknown pathogens (HIV entered the blood supply in the early 1980s).

On the other hand, some risks are avoidable. George Crile continued to transfuse blood into people without determining blood types well after blood typing had become accepted practice. Norman Shumway's heart-transplant patients lived longer than patients at other transplant centers. This information was available to any patients who were willing to look hard enough.

One aspect of risk, however, hasn't yet been presented: nationalism. Although this might seem far-fetched, the issue of nationalism raised its head at the end of 2020, when

several countries participated in the race to develop a coronavirus vaccine. China began to immunize its citizens
with an inactivated viral vaccine, which is made by taking
SARS-CoV-2 virus, growing it in the laboratory, purifying
it, and completely killing it with a chemical. Russia claimed
it would soon immunize its citizens using a strategy called
replication-defective human adenoviruses, made by taking a
virus that causes the common cold (adenovirus), altering it
so that it couldn't possibly reproduce itself (hence the term
"replication-defective"), and genetically engineering it so
that it could make the SARS-CoV-2 spike protein (which is
responsible for attaching the virus to cells). People injected
with this vaccine would make the coronavirus spike protein and then make antibodies to the protein. The United
Kingdom was on the verge of immunizing its citizens with
a vaccine similar to that invented in Russia, except instead
of using a human adenovirus, it used an adenovirus from
monkeys (called replication-defective simian adenoviruses).
And the United States pursued yet a fourth strategy: taking
a small piece of the SARS-CoV-2 gene that coded for the
spike protein (called messenger RNA or mRNA) and directly injecting that little gene into the muscles of its citizens.
Many wondered what would happen if one vaccine worked
much better than another. Would countries switch, or would
they continue to push the vaccine to which they had committed billions of dollars in research funding, development,
and mass production, arguing that their vaccine was good
enough. World leaders also feared that the country with the
best vaccine would restrict use to its citizens and its allies.
Although this scenario might seem unlikely, it's exactly what
happened in the early days of anesthesia, when nationalism
prevailed. As a consequence, European citizens were forced

to use an anesthetic that was far more deadly than the ones used in America.

———

Gore Vidal was an American icon. Before he died in 2012, Vidal wrote more than two dozen novels and numerous plays, essays, and television dramas. He appeared in animated form on *The Simpsons* and *Family Guy* and was a regular on *The Tonight Show Starring Johnny Carson*. Several of his books—such as *Lincoln*, *Burr*, and *1876*—described life in nineteenth-century America. During one press conference, a reporter asked Vidal whether he would have wanted to have lived during that period. "I would never want to live at a time," replied Vidal, "that hadn't perfected anesthesia."

Before anesthesia, surgeons removed bladder stones, drained ovarian cysts, and amputated legs, but little else; they were rewarded for their speed more than their skill. A surgeon named Robert Liston, in an attempt to best his own speed record for amputating a leg, accidentally cut off one of his patient's testicles and two of his assistant's fingers. In the early 1800s, John Collins Warren, a professor at Harvard's famed Massachusetts General Hospital, described how he instructed patients before surgery. "It was the custom to bring the patient into the operation room and place him upon the table," recalled Warren.

> The surgeon would stand with both hands behind his back and say to the patient, "Will you have your leg off or will you not have it off?" If the patient lost courage and said, "No," he had decided not to have the leg amputated, he was carried back to the bed in the ward. If, however, he said, "Yes," he was immediately taken

firmly in hand by a number of strong assistants and the operation went on regardless of whatever he might say thereafter. If his courage failed him after this crucial moment, it was too late, and no attention was paid to his cries of protest. It was found to be the only practicable method by which an operation could be performed under the gruesome conditions which prevailed before the advent of anesthesia.

Perhaps the best description of life before anesthesia came from Fanny Burney who, like Gore Vidal, was a socially prominent author. She was best known for her novel *Evelina, or the History of a Young Lady's Entrance into the World*, which was published in 1778. On September 30, 1811, Dominique-Jean Larrey, a military surgeon who had served under Napoleon, removed a cancerous breast from Burney, who refused to be held down by the seven male attendants and one female nurse who had been brought in to assist with the procedure:

When the dreadful steel was plunged into my breast—cutting through the veins, arteries, flesh, and nerves—I needed no injunction to restrain my cries. I began a scream that lasted unremittingly during the whole time of the incision. And I almost marvel that it does not ring in my ears still, so excruciating was the agony! When the wound was made, and the instrument withdrawn, the pain seemed undiminished for the air that suddenly rushed into those delicate parts and felt like a mass of minute but sharp and forked poniards [daggers] that were tearing at the edges of the wound. Afterward, [the surgeon] was pale, nearly as me, his face streaked with blood and its expression depicting grief, apprehension, and almost horror.

Fanny Burney didn't have to suffer. An anesthetic that would have allowed her to sleep peacefully through the procedure had been available for more than ten years.

=====

Humphry Davy was the son of a middle-class family in Cornwall that cared little about his education. When he was sixteen years old, he went to work for a local apothecary, eventually climbing the ranks to head England's Pneumatic Institution, which used inhaled gases to treat tuberculosis. One of his first experiments was with a gas that had been discovered in 1772 by Joseph Priestley: nitrous oxide. Priestley had made nitrous oxide by simply heating ammonium nitrate, which decomposed to nitrous oxide and water. (He had discovered oxygen the year before.)

First, Davy tested nitrous oxide on animals. When he noticed one of the guinea pigs laughing, he called it "laughing gas." In 1775, Davy tried it on himself. "The thrilling was very rapidly produced," he recalled. "The pleasurable sensation was at first local, and received in the lips, and about the cheeks. It gradually, however, diffused itself over the whole body, and in the middle of the experiment, was as intense and pure as to absorb existence. At this moment and not before, I lost consciousness; it was, however, quickly restored." Davy asked his friends to try it—specifically, poets Samuel Coleridge and Robert Southey and a medical student named Peter Mark Roget, who would later be known for his thesaurus. All agreed that nitrous oxide had two defining characteristics: laughing and unconsciousness.

On January 27, 1800, Davy published a book with the daunting title *Researches, Chemical and Philosophical; Chiefly Concerning Nitrous Oxide, or Dephlogisticated Nitrous Air, and Its Respiration.* On page 556, he wrote, "As nitrous oxide in its extensive operation appears capable of destroying pain, it may probably be used with advantage during surgical operations." In 1812, Humphry Davy was knighted. In 1820, he became president of the Royal Society of London. Unfortunately, his most important contribution to medicine—which was buried in the middle of a long, largely unread and unreadable book—was ignored for almost fifty years.

===

Thirty years after Davy's discovery, nitrous oxide debuted in America—as a carnival sideshow. In 1833, the *Albany Microscope* printed a rave review of a laughing-gas demonstration by a "Dr. Coult of London, New York, and Calcutta." "Dr. Coult," however, wasn't a doctor and he wasn't from London, New York, or Calcutta. He was a nineteen-year-old boy from Connecticut who had dropped out of high school to become a merchant marine, quit his father's textile business, and was now trying to fund a patent for something he had recently invented. To raise money, "Dr. Coult," who traveled around the country in a horse-drawn cart, charged 25 cents to take a sniff of laughing gas, staging two or three shows a day. "The effect which the gas produces upon the system is truly astonishing," wrote the *Microscope*. "The person who inhales it becomes completely insensible and remains in that state for about the space of three minutes, when his senses become restored." People on stage would take a few whiffs and fall down laughing and giggling as if they were drunk. Some passed out completely for minutes at a time. "Dr. Coult" eventually earned the money he needed to patent his handheld revolver. His real name was Samuel Colt.

The carnival barker who changed the course of anesthesia history, however, wasn't Samuel Colt. That honor belonged to a similarly named huckster: Gardner Colton. On

December 10, 1844, Colton staged a show in front of four thousand people at the cavernous Broadway Tabernacle Theater in Hartford, Connecticut. In attendance was a dentist named Horace Wells, who couldn't resist the pitchman standing outside:

> HURRY, HURRY, HURRY! See the great exhibition of the gas that makes you laugh! Feel the pleasurable sensations that rouse the risibilities. This MARVELOUS VAPOR excites every fiber of the body to action and sharpens all the faculties of the mind. STEP RIGHT UP, GET YOUR TICKETS HERE! YE-E-E-S, this gas induces exhilaration, causes uproarious outbursts of UNCONTROLLABLE MIRTH. You, too, can enjoy the delights of intoxication without any of its DEGRADING AFTEREFFECTS! Better than drinking fermented liquors. THRILLING sensations, sublime emotions, exquisite pleasure! Feel lighter than the ATMOSPHERE. Float away into the air! Hey, buy your tickets and go inside! Hey, show ready to start in a few moments. Only a quarter to gain admission! Twenty-five cents, the fourth part of a dollar! STEP RIGHT UP! Don't be afraid to take your lady, Sir. This is a gentlemanly exhibition! You don't have to wait!

Wells climbed onto the stage with his two friends Samuel Cooley and David Clarke, inhaled the gas, and proceeded to make a complete fool of himself. When he got back to his seat, he noticed that Cooley had blood on his pants. "You must have hurt yourself," said Wells. "No," replied Cooley. Wells then said to Clarke that a person could probably "have a tooth extracted or a limb amputated and not feel any pain."

When the show ended, Wells sought out Gardner Colton and asked him to bring a bag of nitrous oxide to his office the next day. On December 11, 1844, Horace Wells became the first person to use nitrous oxide as an anesthetic. Initially, he tried it on himself. Wells asked a fellow dentist, John Riggs, to pull one of his teeth. When he awoke, Wells was amazed that the procedure had been painless. "It is the greatest discovery ever made!" he declared. "I didn't feel as much as the prick of a pin!" During the next few weeks, Wells pulled the teeth of fifteen of his patients using nitrous oxide. None experienced pain. Now, all that Wells had to do was convince the medical community of the value of this miraculous drug.

By January 1845, Wells was ready. To assure the largest, most influential audience, he contacted John Collins Warren, the professor of surgery at Massachusetts General Hospital, and asked for a demonstration. Warren agreed. In February 1845, Wells and Warren stood in the center of an amphitheater surrounded by Harvard medical students and faculty. Wells administered the nitrous oxide and proceeded to pull the patient's tooth. Unfortunately, he hadn't given an adequate amount of the anesthetic. Although the patient initially fell asleep, once the extraction began he screamed out, to peals of laughter from the audience, some of whom shouted, "Humbug!" Wells left the building in disgrace.

Horace Wells's failed experiment left the door open for another dentist, now hailed as the father of anesthesia, even though he isn't.

—

William Morton was born in Charlton, Massachusetts, and studied dentistry in Baltimore before moving to Farmington, Connecticut, where he joined the practice of Horace Wells. Morton wasn't interested in nitrous oxide. Under the advice of a chemist, inventor, and mentor named Charles Jackson,

Morton chose ether instead, which Jackson had assured him was both safer and better. Like nitrous oxide, ether was readily available, easy to make (by mixing sulfuric acid with ethyl alcohol), and had been around since the mid-1800s. And, like nitrous oxide, which was used during "laughing gas parties," ether had its own form of recreational entertainment called "ether frolics."

On September 30, 1846, after testing ether on himself, his dog, and his assistant, Morton painlessly removed a tooth from one of his patients. When the procedure was over, Morton asked, "Are you ready to have the tooth out?" "I am ready," said the patient. "Well, it is out now!" said Morton, showing the patient the newly extracted, bloody tooth. Morton leaked the news to the press. A reporter from the *Boston Journal* wrote, "Last evening, as we were informed by a gentleman who witnessed the operation, an ulcerated tooth was extracted from the mouth of an individual, without giving him the slightest pain. He was put into a kind of sleep by inhaling a preparation, the effects of which lasted about three quarters of a minute, just long enough to extract the tooth." Emboldened, Morton did what Horace Wells had done. He asked John Collins Warren if he could have a public display of his new anesthetic at Massachusetts General Hospital.

On October 16, 1846, in the same amphitheater where Wells had been humiliated a year earlier, ether made its medical debut. Morton began the procedure by administering ether on a sea sponge to Gilbert Abbott, a twenty-year-old housepainter with a huge tumor on his jaw. After three or four minutes, he sank into a deep sleep. Warren then carefully removed the tumor from Abbott's jaw, slowly cutting through bone and muscle. Abbott didn't feel a thing. When the operation was over, Warren looked at the audience and declared, "Gentlemen—*this* is no humbug." After a brief, stunned pause, the students cheered. Morton was a hero.

Oliver Wendell Holmes, a professor of anatomy and the father of the Supreme Court justice, declared Ether Day to be a seminal event in the history of medicine: the birth of modern surgery. Never given to understatement, Holmes said, "By this priceless gift to humanity, the fierce extremity of suffering has been steeped in the waters of forgetfulness, and the deepest furrow in the knotted brow of agony has been smoothed forever." The amphitheater where this event took place still stands at Massachusetts General. It's called the Ether Dome—a monument to the birth of anesthesia in America.

———

William Morton has been hailed as the father of anesthesia because he was the first to administer ether during a public demonstration. But he wasn't the first to use it.

In January 1842, three years before Morton stepped into that amphitheater at Massachusetts General Hospital, Dr. William Clarke of Rochester, New York, administered ether on a soaked towel to a Miss Hobbie while a dentist, Elijah Pope, extracted her tooth. Clarke had become familiar with the chemical during his ether frolics as a medical student.

Two months later, on March 30, 1842, Dr. Crawford Long of Jefferson, Georgia, administered ether to James Venable before removing a tumor from his neck. Long, a graduate of the University of Pennsylvania School of Medicine, was twenty-seven years old at the time. Like William Clarke, Long was familiar with ether, having once bruised himself during an ether frolic without pain. "The ether was given to Mr. Venable on a towel," wrote Long. "And when fully under the influence, I extirpated the tumor. It was encysted, and about a half inch in diameter. The patient continued to inhale ether during the time of the operation; and when informed it was over, seemed incredulous until the tumor had been shown him." Nine months later, on December 17, 1845, Long became the first doctor to anesthetize a woman during childbirth—specifically, his wife during the birth of their daughter.

Neither Crawford Long nor William Clarke realized the significance of what they had done. Long published his experience with ether as a surgical anesthetic in 1849, five years after William Morton had administered it to Gilbert Abbott. Clarke's use of ether during a dental procedure wasn't mentioned in a medical textbook until 1881. Although not considered to be the father of surgical anesthesia, Long would later be recognized with a statue in the US Capitol in Washington, DC.

═══

With the exception of Crawford Long—who died following a massive stroke at the age of sixty-two while handing a newborn baby to an attendant—all of the men involved in the early days of anesthesia met unfortunate ends.

Horace Wells left Massachusetts General Hospital disgraced by the failure of his nitrous oxide demonstration. Depressed, he abandoned his wife and children and moved to New York City, where he became addicted to chloroform. On the evening of January 22, 1848, Wells threw sulfuric acid onto a group of prostitutes, some of whom suffered severe burns on their face. While in jail, Wells killed himself, a handkerchief soaked with chloroform hanging from his mouth. He was thirty-three years old. Julie Fenster, author of *Ether Day: The Strange Tale of America's Greatest Medical Discovery and the Haunted Men Who Made It*, writes, "The man who died in the Tombs [of New York] with a silk scarf stuffed into his mouth and a hat on his head, staring straight into the air, was a chloroform addict. But in Connecticut, Wells was a great man. To apologize for the confusion between the two, he proved himself willing to die."

William Morton spent the rest of his life trying to cash in on his fame by patenting a concoction of ether he had invented called Letheon, which consisted of sulfuric ether plus oil of orange to disguise the sulfur smell. On October 27, 1846, he submitted his invention to the US Patent Office, which refused to recognize it. For twenty years Morton stomped around, railing against a medical establishment that had failed to pay him the fortune he believed he deserved. Eventually, he was censured by the American Medical Association (AMA). In 1868, while taking a carriage ride through New York's Central Park, Morton leaped from his carriage, plunged his head into a nearby lake, and fell unconscious. He was taken to St. Luke's Hospital, where he died a few hours later—the second suicide of the men at the dawn of anesthesia.

Charles Jackson, the chemist who had advised William Morton to use ether instead of nitrous oxide during surgery, also met a tragic end. He had spent much of his life trying to convince his colleagues that he, not Morton, should have been considered the father of modern anesthesia. Of interest, Jackson had also had a lifelong feud with Samuel Morse, who he claimed had stolen his idea for how to communicate over long distances using a telegraph machine. In another attempt to claim his place in history, Jackson administered ether to patients at McLean Asylum for the Insane in Belmont, Massachusetts, believing that the anesthetic could cure insanity. Eventually, Jackson descended into madness and was himself committed to the McLean asylum, where he spent the last seven years of his life.

—

While doctors in the United States argued about the relative merits of ether and nitrous oxide, European doctors chose a third anesthetic—one that would later kill Horace Wells. In the end, only one of the three would survive into the modern era.

James Young Simpson studied medicine in Edinburgh, Scotland, graduating in 1832. By the mid-1840s, Simpson had climbed the ranks to become a professor of midwifery in Edinburgh, relieving the pain of childbirth with ether, like his American colleagues. But Simpson wasn't satisfied. He wanted a more potent agent, one that was pleasant to inhale, worked quicker, and didn't cause vomiting upon awakening. He settled on chloroform, a combination of hydrogen, carbon, and chlorine.

On November 4, 1847, Simpson invited two of his assistants, James Duncan and George Keith, and some of his friends, including a Ms. Petrie, to a dinner party. When the dinner was over, he asked his guests to sniff a variety of volatile gases, including chloroform. Duncan and Keith immediately lost consciousness, falling under the table. Ms. Petrie also lost consciousness, but not before declaring, "I'm an angel! I'm an angel! Oh, I'm an angel!"

The next day, without animal studies, clinical trials, or federal approval, Simpson administered chloroform to a woman during a particularly painful delivery. "I placed her under the influence of chloroform," recalled Simpson, "by moistening half a teaspoon of the liquid onto a pocket handkerchief [and placing it] over her mouth and nostrils. The child was expelled in about twenty minutes. When she awoke, [the mother] observed to me that she had enjoyed a very comfortable sleep." The parents were so elated that they named their daughter Anesthesia. On November 10, 1847, Simpson told a group of colleagues what he had done. Ten days later, he described his experience in a medical journal, claiming that chloroform was more potent and easier to administer than nitrous oxide, and quicker to induce unconsciousness and less flammable than ether. Now the entire medical world knew about it.

James Young Simpson became a national hero. On April 7, 1853, England's leading anesthetist, John Snow, who had been using only ether up to that point, followed Simpson's lead and administered chloroform to Queen Victoria during the birth of Prince Leopold and, four years later, Princess Beatrice. "We are going to have this baby and we are going to have chloroform," insisted the queen. In 1866, for his work on chloroform, Simpson was knighted. His coat of arms read, "Victory over Pain." (Although John Snow was a physician to the queen of England, he is probably best known as the man who proved that contaminated water was the source of a cholera epidemic in London in 1853 that killed ten thousand people.)

It didn't take long, however, for the dangers of chloroform to become apparent. On January 28, 1848, two months after Simpson had described his experiences with chloroform, fifteen-year-old Hannah Greener had a toenail removed. The surgeon, Mr. Meggison, poured chloroform on a cloth and held it near Hannah's nose. "I told her to draw her breath naturally," recalled Meggison, "in about half a minute I observed muscles of the arm become rigid and her breathing a little quickened." Three minutes later, Hannah Greener was dead. An autopsy showed massive congestion of blood in her lungs, probably due to a damaged heart. Chloroform had affected her heart's ability to beat properly.

Hannah's death was the first of many. By the end of 1848, doctors reported six more chloroform deaths. Over the next ten years, fifty more were reported, including a young woman in Boulogne, another woman in Hyderabad, two young men in Lyon, an Australian visiting relatives in Scotland, an Irish woman in the US Navy, a laborer in Westminster, an artilleryman aboard a ship in Mauritius, a patient in Stockholm, a boy from the Scottish Highlands, and various other cases in hospitals in England, France, and Germany. Many of these deaths, similar to that of Hannah Greener's, followed minor operations. By 1863, more than one hundred additional deaths from chloroform had been reported.

In the second half of the nineteenth century, doctors had three choices for anesthesia: nitrous oxide, ether, and chloroform. Each had its drawbacks.

Ether was remarkably safe; it was virtually impossible to die under its influence. The problem with ether, apart from the fact that it had a disagreeable smell, was dramatized for today's television audiences during an episode of *The Knick*, the two-season program on Showtime describing life in a fictional New York City hospital in 1906. In one episode, a surgeon (played by Ben Livingston), while drilling a hole into the front of the windpipe (tracheostomy), catches fire and dies. His face is alight in flames. It is a dramatic scene. And to many who watched it, probably unbelievable. But deaths from ether-associated fires in the operating room were fairly common. The following descriptions from physicians in the mid-1800s reveal the problem with using a highly flammable, volatile gas at a time when fire was the only source of light.

- "While operating at night on a mutilated finger, the lamp being three feet distant, and a sponge placed over the patient's mouth, the air in the vicinity became saturated with ether, ignited, setting fire to the sponge, bed clothes, and even the face of the patient."

- "I used a candle to throw a better light into the mouth. This is held by the assistant not nearer than half a yard from the seat of operation. And now comes the terrible scene. I had scarcely used the ether for twenty seconds when suddenly a volume of flame rushed from the patient's mouth, enveloping the three of us for a single instant."

- "A candle was held by a nurse at a distance of about two feet from the patient's mouth when the vapour suddenly ignited and a scene presented itself that neither I nor any that witnessed it are likely to forget. The man appeared literally to vomit forth fire, while his head seemed, and indeed was, completely enveloped in brilliant flame."

Nitrous oxide, on the other hand, wasn't highly flammable and didn't cause the nausea and vomiting of ether. The problem with nitrous oxide was that it was a relatively weak anesthetic. Patients had to receive virtually 100 percent of the gas to remain unconscious. This meant that they weren't receiving any oxygen during that time. As a consequence,

nitrous oxide couldn't be used during long surgical procedures, for fear of permanent brain damage.

Although ether and nitrous oxide had their limitations,
no anesthetic used in the 1800s and early 1900s was more
dangerous than chloroform, which caused heart damage and
sudden death. For this reason, physicians in the United States
rarely used it. Indeed, physicians in Boston and Philadelphia
had banned it. So, why did Europeans use chloroform for decades when it was clearly the riskier choice? The most likely
explanation is nationalistic pride; chloroform had debuted
in Europe, whereas nitrous oxide and ether had debuted in
America. Also, the births of Prince Leopold and Princess
Beatrice under chloroform anesthesia had been highly and
proudly publicized, influencing a nation. Between 1865 and
1920, after John Snow had clearly proven that ether was the
safer choice, chloroform was used in 80 to 90 percent of all
surgeries in Europe but only rarely in America. Deaths from
chloroform in Europe continued to mount. All of these chloroform deaths were entirely preventable.

———

Of interest, apart from its capacity to kill patients quickly,
chloroform had another problem: it was a favorite among
rapists, murderers, and thieves. In October 1862, Mary Tra-
vers was allegedly raped by an eye surgeon named William
Wilde while under chloroform anesthesia. Wilde would later
become the father of Oscar Wilde. In 1891, Dr. Thomas Neill
Cream was convicted of the murders of at least six women by
chloroform—the first doctor to be convicted of such a crime.
In 1893, during the World's Fair in Chicago, H. H. Holmes,
the first modern-day serial killer, was alleged to have committed at least twenty-seven murders and perhaps as many as

two hundred using chloroform. In September 1900, chloroform was implicated in the death of William Marsh Rice, the
founder of Rice University. In February 1994, a fifty-nine-
year-old British physician was charged with abducting and
raping a seventeen-year-old girl using chloroform. In 2007,
a Cincinnati man confessed to using stun guns containing
chloroform to sexually assault minors.

Despite clear evidence that chloroform was more likely
to kill patients than nitrous oxide or ether, James Young
Simpson continued to promote it until the end of World War
I; after that, ether emerged as the perennial choice in Europe.
But like chloroform, ether, too, would bow out. The last time
it was used at Massachusetts General Hospital was on December 20, 1979. By the early 1980s, its use in America had
disappeared altogether. Only nitrous oxide—which is occasionally used by dentists in addition to local anesthetics—
remains, achieving the unlikely feat of being both the first
and last anesthetic among the three to be used.

———

Although ether, nitrous oxide, and chloroform were all used
recreationally, no anesthetic posed a greater opportunity for
addiction than the first local anesthetic.

For centuries, it was known that those who chewed
leaves from the coca plant could numb their mouth, lips, and
tongue. In 1858, Albert Niemann successfully purified this
substance, calling it cocaine. The use of cocaine as a medical
product wasn't realized until Carl Koller, a young Viennese
ophthalmologist, put a small amount of it on his tongue and
marveled at the completeness of its numbing properties.

Koller tested cocaine by putting a liquid form of the
drug onto the eye of a frog. First, he touched the animal's

cornea. No response. Then he scratched its cornea with a needle. Again, no response. Koller applied an electric current to
the cornea; then he cauterized it with silver nitrate. Nothing,
it appeared, caused the frog, which was fully awake at the
time, to twitch. Then Koller took the next step. He put liquid
cocaine onto his own eye and touched his cornea with a pin.
Like the frog, Koller felt nothing.

On September 11, 1884, Carl Koller successfully oper-
ated on the cornea of a patient with glaucoma using cocaine.
Local anesthesia was born. No longer did some patients
have to suffer the possible harms of general anesthesia when
blocking pain locally would do. That same year, a pharmaceutical company founded in Darmstadt, Germany, by Friedrich Jacob Merck distributed cocaine to physicians for test
purposes. One of those physicians was a good friend of Carl
Koller's, a man who had graduated from the same medical
school in Vienna: Sigmund Freud.

Freud introduced cocaine into medicine by giving it to
a man who suffered from persistent, excruciating pain in his
left thumb following an injury. At the time, his patient was
addicted to morphine. Freud offered cocaine instead. With
Freud's inadvertent help, the man became a cocaine addict.
Freud had become the first physician to use cocaine for something other than surgery, and he continued to convert morphine addicts to cocaine addicts until his death.

The most famous physician to promote the use of cocaine as a local anesthetic was William Halsted, one of the
founders of Johns Hopkins Hospital in Baltimore. Halsted,
who had been the first to use rubber gloves in an operating
room, experimented with cocaine as a nerve block for regional
and local anesthesia. Unfortunately, he became hopelessly
addicted to the drug. (In *The Knick*, the principal character,

Dr. John W. Thackery, is based on the life of William Halsted. Played by Clive Owen, Thackery, too, is a cocaine addict.) At the time, cocaine's addictive properties hadn't been recognized. But when this became obvious, it ended the medical use of the drug, later to be replaced by nonaddictive local anesthetics such as lidocaine in the 1940s.

—

The twentieth century witnessed a revolution in general anesthesia. In 1929, cyclopropane was invented. Like nitrous oxide, ether, and chloroform, cyclopropane was inhaled as a gas, but it wasn't accompanied by their unfortunate side effects. That same year, sodium amytal, the first intravenous anesthetic, was invented. This was followed by a parade of other critical inventions. General anesthetics such as hexobarbital in 1932, thiopental in 1934, fluroxene in 1952, halothane in 1958, and eventually a raft of other fluorinated hydrocarbons such as methoxyflurane in 1960, enflurane in 1971, sevoflurane in 1975, isoflurane in 1979, and desflurane in 1992. Each made for safer, better anesthesia. Muscle relaxants, such as curare in 1939 and succinylcholine in 1949, made it easier for anesthesiologists to control unwanted muscle movements during surgery. Breathing tubes and breathing machines like the positive-pressure ventilator, invented in 1952, allowed anesthesiologists to control a patient's breathing. Short-acting anesthetics like propofol in 1977 also allowed anesthesiologists to induce rapid, short-lived anesthesia.

Horace Wells, William Morton, Charles Jackson, John Collins Warren, James Young Simpson, and John Snow wouldn't recognize anesthesia today. A single anesthetic agent inhaled on a soaked rag or sea sponge has been replaced by needles and a panoply of drugs whose functions include analgesia, anesthesia, amnesia, and muscle relaxation. Doctors keeping their fingers on a patient's pulse during surgery have been replaced by beeping and flashing monitors showing the patient's heart rate and rhythm, respiratory rate, and blood pressure. Nonetheless, the possibility of death from general anesthesia remains.

On April 22, 1982, ABC's 20/20 aired a program titled "The Deep Sleep." The show began with an ominous warning: "If you are going to go into anesthesia, you are going on a long trip and you should not do it if you can avoid it in any way. General anesthesia is safe most of the time, but there are dangers from human error, carelessness, and a critical shortage of anesthesiologists." The program claimed that among the twenty million surgeries that were performed every year, six thousand Americans would die or suffer permanent brain damage from general anesthesia. If the ABC program is correct, this means that the rate of death or permanent harm from anesthesia at the time was about one in 3,333. Today, with advances in general anesthetic agents and complementary drugs, the rate of death or permanent harm following surgery is about one in one hundred thousand. Which is much better than one in 3,333, but still not zero. The perfect anesthetic agent has yet to be invented. Perhaps most surprising is that, despite anesthetics' availability for more than 150 years, the exact mechanism by which they work remains elusive.

———

Before leaving the section on risk, here's one more way to think about it: Imagine that you are walking through the woods and come upon a steep, narrow gorge. A bridge over the gorge will allow you to continue. You know that the bridge is old and in a state of disrepair. But you also know that millions of people have walked across it without a problem.

This analogy applies to several medical decisions: People choose general anesthesia during surgery, even though every year one of every one hundred thousand people in the United States will die from the anesthetic. People choose antibiotics such as sulfanilamide to treat a variety of bacterial infections, even though three of every one hundred thousand people who receive them will suffer severe and occasionally fatal allergic reactions. People choose vaccines, even though about one of every 1.4 million doses is complicated by a severe, immediate allergic reaction that requires treatment with a shot of epinephrine.

Most people make these choices unaware of these known risks. But whether they know about the risks or not, the benefits of anesthesia, antibiotics, and vaccines clearly outweigh them. So, the choices are easy.

In a second scenario, you approach the gorge knowing that the bridge is old and in a state of disrepair but not knowing whether it can safely carry you across. Few if any people have crossed the bridge before. You also know that there is a different bridge about three miles up the path that is safe and has been traveled recently by others.

This analogy applies to a second group of medical decisions. For example, people who require frequent transfusions with blood or blood products (like those with sickle cell disease or hemophilia) might choose to wait until more people have received artificial blood before trying it on themselves or their children. For now, they'll walk to the next bridge—the one that has been crossed before—even though they remain at constant risk of receiving blood containing a pathogen, known or unknown, that hasn't been included in the screening test.

In a third scenario, you approach the gorge knowing that the bridge is old and in a state of disrepair and not knowing whether it can safely carry you across. Now, however, a lion is chasing you. There's no time to walk to the other bridge.

This analogy applies to a third group of medical decisions. Louis Washkansky and Philip Blaiberg, as noted by their surgeon, Christiaan Barnard, were being chased by a lion. If they didn't get a heart transplant, which at least offered the possibility of a longer life, they would soon die. Today, people on a heart-transplant waiting list who are unlikely to live much longer might choose to be among the first to receive a pig's heart because they fear they may be among the 1,300 people on the list every year who die while waiting.

In the final analysis, people should make the most informed, most clear-eyed, most dispassionate decisions using all of the information available, knowing that decisions under uncertainty might be the wrong ones. Every choice involves risk, even the simplest ones.

PART II

OVERSIGHT

IN 2020, AN ELECTION YEAR, SEVERAL FORCES WERE INFLUENCING THE development of COVID-19 vaccines. President Donald Trump promised a COVID-19 vaccine by election day, November 3, even though it was virtually impossible to complete the large clinical trials, which included more than thirty thousand participants for each vaccine, by that time. Trump also vowed to "bypass the FDA" if the vaccine wasn't released in a timely manner. The Trump administration had previously imposed its will on several science-based agencies. It had insisted that the Environmental Protection Agency remove the phrase "climate change" from its website; that the National Oceanic and Atmospheric Administration, which oversees the National Weather Service, support the administration's claim that Hurricane Dorian had hit Alabama when it hadn't; and that the FDA approve the use of a drug, hydroxychloroquine, for the treatment of COVID-19, even though it had never been shown to work and was known to cause heart problems. In response to the administration's pressure, the FDA had rapidly approved hydroxychloroquine through the emergency use authorization—the same lightning-quick mechanism by which COVID-19 vaccines were about to be approved.

In response to the fear that COVID-19 vaccines were about to be released without adequate FDA oversight, the House Committee on Energy and Commerce, the House Committee on Oversight and Reform, and the Senate Committee on Health, Education, Labor, and Pensions held hearings insisting that the FDA commissioner, Dr. Stephen Hahn, resist pressure from the administration to release vaccines before they had been adequately tested. (I had to testify at one of those hearings.) A bill was then put before Congress calling on the FDA to rely on the recommendations of its vaccine advisory committee before granting approval. Similarly, a letter signed by hundreds of prominent researchers pleaded with Commissioner Hahn to do the right thing.

In response, Commissioner Hahn wrote opinion pieces in the *Washington Post* and the *Journal of the American Medical Association* insisting that he would approve COVID-19 vaccines only after they had been shown to be safe and effective and only after they had been approved by the FDA vaccine advisory committee. (I was on that committee.) While Hahn's statements were reassuring, many people inside and outside the administration continued to worry about how all this was going to play out.

The FDA's willingness to approve hydroxychloroquine, which didn't work to treat or prevent COVID-19 and was unsafe, and later to approve convalescent plasma, which also hadn't been shown to work, caused many in the press and the public to worry that the FDA was serving at the will of the administration instead of protecting Americans from pharmaceutical products that might be unsafe or ineffective.

The public no longer trusted the FDA. As a consequence, chaos ruled. By the end of 2020, the National Medical

Association, an organization composed of African American physicians, said that it would form its own committee to advise its members, ignoring the recommendations of the FDA. Several states followed suit, saying that they, too, would form their own expert committees. The mistrust of the FDA reached its illogical end when several pharmaceutical companies wrote an open letter to the American public, stating that they wouldn't release any vaccine until it had been shown to be safe and effective. In essence, the companies were saying, "You can trust us, even though we know that you don't trust the FDA to regulate us."

The fear surrounding the possible lack of federal oversight of COVID-19 vaccines was motivated by one thing: history. The purpose of oversight by federal agencies or professional societies is to avoid the errors, misrepresentations, willful ignorance, greed, and fraud that are far too often part of the human endeavor. Indeed, every law regulating the pharmaceutical industry was, as you'll see in the chapters that follow, in response to a major tragedy. "The history of drug regulation," writes historian Michael Harris, "is written on tombstones." If Americans had gotten to the point that they no longer trusted the FDA to regulate the pharmaceutical industry, our country was in trouble.

CHAPTER 4

"JIM"

Biologicals

Sunday, October 19, 1901

"I was called in on October 19 by Mrs. Baker to attend her daughter, Bessie, who was suffering a severe case of diphtheria," recalled Dr. R. C. Harris of Saint Louis, Missouri. "I injected the usual quantity of antitoxin into the little girl." Diphtheria antitoxin, which had been available for several years, had saved the lives of diphtheria sufferers. Its value was unquestioned. Harris injected the antiserum into Bessie's sister and brother. "As a precautionary measure, frequently adopted in such cases, I also injected a quantity of antitoxin in the two smaller children," said Harris. "In three or four days a marked improvement was apparent in the case of the little girl, Bessie, and I concluded that she was out of danger and would soon be entirely well."

Sunday, October 26

One week later, the situation in the Baker home had taken a turn for the worse. "I was hurriedly called to Mrs. Baker's home over the confectionary," said Harris. "There I found the little girl was suffering from tetanus. I could do nothing for her. The poison was injected so thoroughly through her system that she was beyond medical aid."

Monday, October 27

"[Bessie] died at noon Monday," lamented Harris. "Before Bessie died, the second child, May, developed symptoms of the terrible disease. I was called again but could do nothing." At midnight, May Baker, Bessie's four-year-old sister, also died from tetanus.

Wednesday, October 29

Frankie, Bessie's two-year-old brother, developed symptoms of tetanus. "He lingered until Tuesday night at midnight," said Harris. "I called there again Wednesday morning and found that the third child had also developed symptoms of the disease. I fear that he will die, too, as his condition is apparently very critical." Frankie, however, survived.

Deaths from tetanus weren't confined to the Baker home. Also on October 29, Dr. M. Gohand reported the death of an eleven-year-old named Jacob Centurio.

By November 7, thirteen children in Saint Louis were dead: Veronica Keenan, age five; Agnes Adele Keenan, age seven; Mamie Keenan, age ten; Frank Novak, age three; Ettie Simon, age five; Ike Stein, age four; Flora Fuerst, age eight; Emma Mary Ernst, age four; Charles Cytron, age eleven; Bessie Baker, age six; May Baker, age four; Jacob Centurio, age eleven; and Nettie Kammerman, age unstated.

These children shared several features. Many had suffered from diphtheria, but none had died from the disease.

Instead, all had died from tetanus. Most worrisome, all had been treated with the same batch of diphtheria antiserum obtained from the same place—the Saint Louis Health Department. And all the antiserum bore the label "August 24, 1901." The source of the antiserum was a horse named "Jim."

———

Early in the twentieth century, diphtheria was a notorious strangler of children. The disease typically spread from one person to another by coughing, sneezing, or even talking. A few days after coming into contact with someone who was infected, symptoms would begin. Mild at first, with irritability, decreased activity, and low-grade fever, the disease quickly progressed to a grotesque enlargement of the lymph glands located under the jaw, causing a frightening "bull-neck" appearance. After the neck swelled, the signature feature of diphtheria appeared: a thick, gray, leather-tough membrane that stuck to the back of the throat, sometimes extending into the windpipe. (The word "diphtheria" is derived from the ancient Greek word meaning "leather.") This membrane was virtually impossible to peel off without ripping the lining of the windpipe. Sometimes doctors were able to save lives by performing a tracheostomy. Most times, however, the sticky membrane completely blocked the airway before a tracheostomy could be performed, no different than being smothered by a pillow. Suffocation wasn't the only way to die. Children also suffered from shock when a toxin made by diphtheria affected the heart muscle or paralysis when it affected the nerves. It wasn't hard for doctors to diagnose diphtheria. Nothing else was like it.

In 1901, when Bessie Baker and the children of Saint Louis suffered diphtheria, antibiotics hadn't been invented yet. But thanks to a series of discoveries by French and

German researchers two decades earlier, a treatment in the form of diphtheria antisera was available.

In 1883, Edwin Klebs isolated a bacterium from the back of the throat of a child with diphtheria. In 1884, Friedrich Loeffler grew the bacteria in the laboratory, calling it *Corynebacterium diphtheriae*. One thing, however, didn't make sense. While Loeffler found diphtheria bacteria in the throat of patients with the disease, he couldn't find them anywhere else. How, then, did diphtheria bacteria affect the heart and nerves if they never left the back of the throat?

In 1888, Émile Roux and Alexandre Yersin figured it out. They identified a poison (toxin) made by the bacterium that traveled from the throat to the rest of the body. When they purified this toxin and injected it into animals, the heart and nerve symptoms appeared, just like in people.

In 1890, a German scientist named Emil von Behring took the next step—one that would soon save the lives of those who were infected. Starting with guinea pigs and moving to dogs, goats, horses, and sheep, Behring found that animals injected with diphtheria toxin produced a substance in the blood (called antitoxin or antiserum) that neutralized its effects. While Behring was performing his studies—between 1886 and 1888—more than forty thousand cases of diphtheria had occurred in the German empire; ten thousand were fatal. By September 1894, after Behring had shown that his antisera could treat diphtheria, the death rate from the disease was cut in half.

On October 30, 1901—the same month and year as the Saint Louis tragedy—the Nobel Committee in Stockholm awarded the first ever Nobel Prize in Physiology or Medicine to Emil von Behring for "his works in relation to serum therapy and especially its use against diphtheria." The presentation, which marked the fifth anniversary of the death of Alfred Nobel, took place in the great room of the Royal Swedish Academy of Music. In attendance were members of the Swedish royal family, government ministers, and leading lights in the fields of literature, science, and the arts. Although the Nobel Prize hadn't reached the level of international acclaim that it has today, every major medical journal—including *Lancet*, *Nature*, *Science*, and *Scientific American*—described its significance. Not one, however, mentioned what had just occurred in Saint Louis.

———

In August 1894, seven years before the Saint Louis tragedy, the German firm Hoechst became the first major pharmaceutical company to make diphtheria antiserum. By 1896, Parke, Davis & Co. of Detroit and H. K. Mulford of Philadelphia also joined in. By the late 1890s, at least five pharmaceutical companies were making diphtheria antisera. Manufacture wasn't limited to pharmaceutical companies. The New York City Health Department made its own antiserum, as did the Saint Louis Health Department. And they all made it the same way.

First, scientists grew diphtheria bacteria in the laboratory. Then they purified the toxin away from the bacteria before injecting it into horses. Antibodies directed against the toxin formed in the horse's bloodstream. Investigators then bled the horse and let the blood clot, leaving only the serum. To test its potency, guinea pigs were injected with diphtheria toxin along with various dilutions of the antiserum. Only high potency antiserum was released to the public. Most importantly, every batch of antitoxin was injected into guinea pigs to make sure that it didn't contain tetanus, a disease known to occur in horses (and every other mammal that walked or crawled on the face of the earth).

——

Like diphtheria, tetanus was also caused by a bacterium (*Clostridium tetani*). Unlike diphtheria, however, tetanus wasn't spread from one person to another. Rather, it was contracted through a scratch, cut, or burn that was contaminated with soil containing tetanus spores. Once the tetanus bacteria entered the body, they, like diphtheria, produced a powerful toxin that caused painful muscle contractions, sometimes making it difficult to open the mouth (which is why the disease is often called lockjaw). In addition, victims suffered seizures, difficulty swallowing, changes in blood pressure and heart rate, and intolerance to light or sound, which would precipitate a series of violent muscle spasms. When tetanus toxin affected the vocal cords or muscles necessary for breathing, patients had a tube inserted into their windpipe and were placed on a breathing machine.

The good news is that tetanus is not a subtle disease. If a horse used to make diphtheria antisera had tetanus, it would be obvious. And any antisera that was made from that horse would immediately be discarded.

——

The Saint Louis antiserum facility, which launched in September 1895, was a low-budget affair consisting of a part-time bacteriologist named Armand Ravold, a "janitor" named Henry Taylor, and a 1,600-pound ambulance horse named Jim that had been retired to the grounds of the poorhouse where the facility was located. During his three years in the facility, Jim had produced more than seven gallons of antitoxin. As a consequence, the death rate from diphtheria in Saint Louis had dropped from 35 percent to 6 percent. On August 24, 1901, two quarts of antiserum were harvested from Jim. Two hundred vials were then distributed to various doctors throughout the city.

On September 30, Ravold bled Jim to obtain a fresh batch of antiserum. Three days later, however, on October 2, Jim became ill with tetanus and was destroyed. It was assumed that the antiserum obtained on September 30 was contaminated with tetanus toxin, so it was discarded without further testing. Fortunately, the children in Saint Louis had received a batch collected on August 24, well before Jim had acquired tetanus. Also, the August 24 antisera had been thoroughly tested for tetanus toxin before release. So it was, for all appearances, perfectly safe.

When the tetanus deaths were first reported, Ravold was inconsolable. "I do not dare think about it," he said. "I have other work to occupy my mind. If I did not, I would go crazy. I have been saving lives almost daily by this work, and my heart and soul were in the effort to save the little ones from suffering and death. Now comes this awful thing—it is horrible. To save my life I could not tell how this happened." He was convinced, however, that the antiserum from August 24 was blameless. After all, it had been collected well before Jim developed symptoms of tetanus, and it had been shown to be free of tetanus toxin after it was injected into guinea pigs. Also, the antisera obtained on September 30, which likely was contaminated with tetanus toxin, had been discarded.

Like Ravold, the commissioner of health in Saint Louis, Dr. Starkloff, was convinced that the August 24 antiserum wasn't responsible for the deaths. He was worried that citizens would now have an unfounded fear of antitoxin. "Whatever may have been the cause of these deaths," he said, "it was no fault of the serum treatment. That is still the most efficacious method of treating diphtheria, and the unthinking,

the ignorant, I greatly fear, will become afraid of the name
antitoxin and physicians [will] have great difficulty in making use of the treatment." Starkloff then explained how little
tetanus toxin was necessary to kill a guinea pig during safety
tests. "[About] 0.00000005 grams," he said. "A quantity too
small for human comprehension." The August 24 antitoxin
had been tested in at least a dozen guinea pigs. Given the sensitivity of the test, it was simply not possible that the August
24 antisera could have contained tetanus toxin.

Starkloff had reason to worry about how the public
would now view diphtheria antitoxin following the deaths.
The Chicago Health Department's weekly bulletin had recently put out a statement that the death rate from diphtheria
had increased by more than 30 percent, all because people
had been frightened by the disaster in Saint Louis.

———

On December 17, 1901, the Tetanus Board of Inquiry in Saint
Louis interviewed Dr. Armand Ravold, who insisted that he
and Henry Taylor—the "colored janitor," as he was described
in the newspaper—had disposed of the September 30 serum
after Jim had died. Both Ravold and Taylor insisted that the
batches of antisera collected on August 24 and September 30
hadn't been mixed. But Taylor's story changed after he was
"closeted with the Chief of Detectives by order of Mayor
Wells." Taylor was put back on the stand, where he admitted
that he had mixed the August 24 antisera with the September
30 antisera because he thought it was safe and because the
supply of the August 24 antisera had almost been exhausted.
In other words, although the antisera had been labeled "August 24," the vials that had been distributed to doctors also
contained the tetanus-contaminated sera from September 30.

On February 13, 1902, members of the Tetanus Board of Inquiry issued their verdict. They concluded that the thirteen children had all died from the antisera collected on September 30; that Henry Taylor was not aware of its poisonous nature; and that Armand Ravold was negligent in ensuring that the tainted antisera was destroyed. Further, Taylor and Ravold were to be dismissed from the Health Department. Ravold defended his coworker. "Taylor, a man of sixty-five, honest and faithful, was not supposed to be competent to look after the professional affairs of the office," said Ravold. "He was simply a good servant . . . and this discharge will leave him in hard times."

Ravold went on to have a distinguished career as a bacteriologist, later becoming president of the Saint Louis Medical Society. He died in 1942 at the age of eighty-three, his obituary making no mention of the tragedy that had occurred forty years earlier. Henry Taylor, over Ravold's objections, was fired from the Saint Louis Health Department. He eventually found work as a waiter and caterer. On June 6, 1907, Taylor died from a kidney condition and cancer. His death certificate was signed by Armand Ravold.

———

The Saint Louis disaster—along with a similar disaster in Camden, New Jersey, where nine children died from a tetanus-contaminated smallpox vaccine—created a national outcry. The public now insisted that companies and health agencies be held to a higher standard of production and testing.

As a consequence, on July 1, 1902, President Theodore Roosevelt signed the Biologics Control Act, also known as the Virus-Toxin Law. Following inspections, which could be unannounced, all manufacturers of vaccines, sera, and antitoxins had to be licensed annually. Product labels had to clearly show the product name, license number, and expiration date, and a qualified scientist had to supervise every aspect of production. Punishments for violations were fines up to $500 and up to a year in prison. The government assigned the task of oversight to the Hygienic Laboratory of the Public Health Service. In 1930, the Hygienic Laboratory changed its name to the National Institute of Health and in 1948, as more institutes were added, to the National Institutes of Health. In 1972, regulation of biologicals was transferred to the Food and Drug Administration.

———

In the 1940s, a vaccine was developed to prevent diphtheria. The vaccine was made by inactivating diphtheria toxin with a chemical. When injected into people, the inactivated toxin induced high levels of antitoxin in the bloodstream. The vaccine allowed people to make their own antitoxin, no longer having to rely on horses to do it for them. As a consequence, diphtheria has been virtually eliminated from the United States. Between 1980 and 2014, only fifty-seven cases were reported. As of January 6, 1997, diphtheria antiserum was no longer commercially available. The antisera, and its inventor, have been largely forgotten—with the exception of one popular story for children.

In January 1925, diphtheria antiserum was rushed seven hundred miles by dogsled from Nenana to Nome, Alaska, just in time to avoid a diphtheria epidemic. By the end of the year, a statue of the lead dog, Balto, was erected in New York City's Central Park. Beginning in 1973, this heroic trip was commemorated with a yearly recreation of the event: the Iditarod. And in 1995, the popular animated movie *Balto* became an instant family favorite.

CHAPTER 5

JOAN MARLAR

Antibiotics

ON SEPTEMBER 26, 1937, SIX-YEAR-OLD JOAN MARLAR WAS TAKEN TO her doctor, Logan Spann, who diagnosed strep throat. Spann prescribed a new, breakthrough antibiotic: sulfanilamide. During the next nine days, Joan suffered nausea, vomiting, and stomach pains; her kidneys shut down, she became frighteningly weak, and she lapsed into a coma. On October 5, at Tulsa's Morningside Hospital, little Joan Marlar died from kidney failure. The doctors said it was from a strep infection of her kidneys. Two months later, on November 8, 1937, when it had become clear what had actually killed her daughter, Maise Nidiffer wrote a letter to President Franklin Delano Roosevelt.

> Dear Sir:
>
> Two months ago, I was happy working taking care of my two little girls, Joan age 6 and Jean age 9. Our byword through the depression was that we had good health and each other. Joan thought her mother was right in everything, and it would have

made your heart feel good last November to have seen her jumping and shouting as we listened to your reelection over the radio.

Tonight, Mr. Roosevelt, that little voice is stilled. The first time I ever had occasion to call a doctor for her and she was given the Elixir of Sulfanilamide. Tonight, our little home is bleak and full of despair. All that is left to us is the caring for of that little grave. Even the memory of her is mixed with sorrow for we can see her little body tossing to and fro and hear that little voice screaming with pain and it seems as though it would drive me insane. During her 9 days of illness as we sat by her bed only once did those little eyes lose their dull and unknowing look. Jean and I begged her to look and to know us. A smile broke over her face and she laughed aloud with us and as quickly it vanished, never to smile and know us again.

Tonight, President Roosevelt, as you enjoy your little grandchildren of whom we read about, it is my plea that you will take steps to prevent such sale of drugs that will take little lives and leave such suffering behind and such a bleak outlook on the future as I have tonight.

In my confidence in you I am writing you and hope that you can realize a little of what I am suffering and that you will take steps to prevent such in the future for I realize also there are other homes where hearts are broken such as mine.

It is easy for people to say, "Try to think that she died that others might live." It is easier to say when it doesn't strike in your own home.

Enclosed is the picture of the baby I grieve for day and night.

Thanking you and sincerely,

(Mrs.) Maise Nidiffer

Eleanor Roosevelt read the letter to her husband. Then she released it to the media. Joan Marlar was among the first victims of a much larger, much graver tragedy.

———

Maise Nidiffer was convinced that Elixir Sulfanilamide had killed her daughter. Surprising, given that only a few years earlier it had been hailed as "the magic bullet" for the treatment of a wide range of bacterial infections.

Sulfanilamide was a product of the German dye industry. The company that made it, IG Farben, was the largest chemical and pharmaceutical company in the world. In the 1930s, IG Farben would achieve worldwide recognition for the Nobel Prize–winning invention of the first antibiotic. Later, the company would receive worldwide condemnation for employing slave labor from concentration camps, including thirty thousand prisoners from Auschwitz, and for the manufacture of the cyanide gas Zyklon-B, used to kill more than one million people during the Holocaust.

The unlikely production of a lifesaving antibiotic by a company that manufactured dyes was based on an idea by Paul Ehrlich, a German physician and scientist. Ehrlich, who had created thousands of dyes, found that some stained only human cells while others stained only bacteria. He reasoned that if he could link a dye that stained only bacteria to a toxic agent, he could kill bacteria without harming human cells. The man who made Ehrlich's dream a reality was Gerhard Domagk.

In 1915, Domagk left the University of Kiel to serve in the German army's medical corps at the Russian front. When the war ended, of the thirty-three university students who had served with him, only three were still alive. In August 1927, Domagk joined a team of researchers at IG Farben.

Inspired by Ehrlich's idea, Domagk, who had been horrified by how little could be done for wound infections during the war, linked a bright red, bacteria-staining dye to an agent, sulfanilamide, that he believed could kill one of the world's most deadly bacteria: streptococcus. It was streptococcus that had been primarily responsible for the fatal wound infections Domagk had witnessed at the Russian front. The final product was called Prontosil, which IG Farben patented on Christmas Day 1932. Domagk found that if he infected mice with streptococcus, they were usually dead in a day or two. But if he treated them with Prontosil, they survived. In 1933, a Düsseldorf physician became the first to prescribe Prontosil to save the life of a child with a severe bloodstream infection (sepsis) caused by streptococcus. For Domagk, the success of Prontosil would soon become even more personal.

On December 4, 1935, while working on a Christmas gift, Domagk's young daughter pierced her hand with a needle. The next day, her hand and the lymph glands under her arm swelled, her temperature rose, and she became dizzy and gravely ill. Domagk took a swab of the pus pouring out of his daughter's hand and found that it was teeming with streptococcus—a ghastly reminder of what he had witnessed during the war. "I asked the permission of the treating surgeon to use Prontosil," recalled Domagk. The next day his daughter's temperature returned to normal. Her life saved.

Americans would learn of the miracle of sulfanilamide the following year. In December 1936, Franklin Delano Roosevelt's son was admitted to Massachusetts General Hospital in Boston with a severe sinus infection that had worsened. In a last-ditch effort to save his life, doctors gave him sulfanilamide. FDR Jr.'s complete recovery was chronicled by the *New York Times* under the headline, "Young Roosevelt Saved by New Drug." *Time* magazine called sulfanilamide "the medical discovery of the decade."

In 1939, Gerhard Domagk was awarded the Nobel Prize in Medicine for his "recognition of the antibacterial activity of Prontosil." Adolf Hitler refused to allow him to travel to Sweden to receive it. Hitler had been angered by the Nobel Prize Committee's decision in 1936 to offer the prize to a German dissident and pacifist named Carl von Ossietzky, whom Hitler had sent to a concentration camp. Hitler had issued a decree that no German would be allowed to receive the prize. Unsure how to proceed, Domagk wrote a letter to the Nobel Committee explaining his dilemma. Unfortunately, he hadn't checked with German authorities before sending it. This was a mistake. On November 17, 1938, Gestapo agents, with weapons drawn, roughly ushered Domagk out of his house and threw him into an isolated cell in the Wuppertal police station. The first night of his captivity, one of the jailors asked him why he was there. "Because I received the Nobel Prize," said Domagk. Later that night, the jailor told another guard, "In that cell we have a crazy man." Hitler successfully forced Domagk to refuse the prize. It was only years later that he received the medal and certificate.

When scientists at IG Farben found that sulfanilamide also worked without linking it to a dye, Prontosil was abandoned in favor of pure sulfanilamide. (Given that the dye in Prontosil occasionally turned people "lobster red," this was a welcome advance.) During the next few years, doctors found that sulfanilamide, in addition to killing streptococcus, killed a wide array of other bacteria, including those that caused pneumonia, gonorrhea, sepsis, meningitis, dysentery, and kidney and bladder infections. By 1942, sulfanilamide was estimated to have saved twenty-five thousand lives per year from pneumonia alone. The death rate from meningitis dropped from 90 percent to 10 percent.

Like any medicine that has a positive effect, sulfanilamide had negative effects. An editorial in the *Journal of the American Medical Association* on July 31, 1937, less than two months before the death of Joan Marlar, warned about the indiscriminate use of the drug, which was often sold over the counter. Sulfanilamide had been shown to cause dermatitis (inflammation of the skin), photosensitization (a rash caused by sensitivity to sunlight), agranulocytosis (a decrease in white blood cells necessary to fight infection), and methemoglobinemia (a dysfunctional form of hemoglobin that is less capable of carrying oxygen to the body and can be deadly). With the death of Joan Marlar, some physicians now wondered about a fatal new side effect of the drug that had gone previously unreported. Was this lifesaving medicine now also causing kidney failure and death?

———

On Saturday, October 9, 1937, four days after Joan Marlar had died, a group of worried physicians sent a telegram to the American Medical Association: "ATTENTION IS DRAWN TO AT LEAST SIX DEATHS FOLLOWING ADMINISTRATION OF ELIXIR SULFANILAMIDE." In the span of ten days, six children had died in the small oil town of Tulsa, Oklahoma. All had strep throat. All were diagnosed initially with kidney failure caused by strep. And all had taken Elixir Sulfanilamide, a product of the S. E. Massengill Company of Bristol, Tennessee. After receiving fewer than three tablespoons of the drug, two-year-old Bobbie Summer suffered vomiting, bloody urine, a distended abdomen, and eventually coma and death. The following day, eleven-month-old Mary Watters and eight-year-old Jack King Jr. died with the same symptoms. Soon, three more children—five-year-old Sonny Wokeford, six-year-old Michael Sheehan, and six-year-old Joan Marlar—were also dead.

In response to the telegram, Dr. Paul Leech, secretary of the AMA's Council on Pharmacy and Chemistry replied: "WOULD APPRECIATE RECEIVING FULL DETAILS CONCERNING DEATHS. IF THIS OFFICE CAN AID YOU FURTHER PLEASE ADVISE." The choice by physicians in Tulsa to send their telegram to the AMA and not the FDA underlines the pathetic state of drug regulation in the United States in 1937. For all practical purposes, drugs were unregulated. The only supervision, if you could call it that, was by the Council on Pharmacy and Chemistry, a division within the AMA that had been formed in 1905 to examine the composition of drugs for sale in the United States. The submission process, however, was entirely voluntary. If they chose, pharmaceutical companies could sell their products without ever meeting any standards of safety, effectiveness, or good manufacturing practices. Companies were also under no obligation to list the contents of their products on the label.

On Sunday, October 10, the day after the group of Tulsa doctors had sent their telegram, Dr. James Stevenson, president of the Tulsa County Medical Society, sent a second telegram to the AMA: "MANY CASES [WITH] COMPLETE ANURIA WITH UREMIA. TULSA COUNTY MEDICAL SOCIETY MEETS MONDAY NIGHT. DESIRE ALL POSSIBLE INFORMATION AT ONCE. WIRE OR PHONE ME MY EXPENSE." The word anuria refers to a patient's inability to produce urine, a sign of kidney failure. Uremia refers to the buildup of toxic products in the bloodstream that happens when kidneys fail. Stevenson

wanted to know what the Council on Pharmacy and Chemistry had found when it had examined Elixir Sulfanilamide.
What exactly was in those bottles? He was suspicious that
the diagnosis of kidney failure caused by strep was wrong.
He worried that what his colleagues were actually wit-
nessing in Tulsa was a mass poisoning caused by mercury.
Stevenson knew that some medical products contained mercury, which had been used to treat syphilis. And he knew
that mercury could cause kidney failure. Leech's response
was disheartening: "NO PRODUCT MASSENGILL COMPANY
ACCEPTED BY COUNCIL ON PHARMACY AND CHEMISTRY."
Massengill had never sent Elixir Sulfanilamide to the AMA
for testing. Nor did it have to. Only Massengill knew what
was in those bottles.

Massengill wasn't the only pharmaceutical company to
make sulfanilamide in the fall of 1937. Merck, E. R. Squibb,
Winthrop Chemical, and Parke, Davis & Co. also made the
drug. These other products, however, were in powder or tab-
let form. Massengill's was a liquid. Sulfanilamide, as it turned
out, was hard to dissolve. Massengill had solved the problem,
making an elixir that was much easier to give to children.
Merck, Squibb, Winthrop, and Parke Davis had all sent their
products to the AMA for analysis before putting them up for
sale. Massengill hadn't. The only way that Stevenson could
have learned the contents of Elixir Sulfanilamide would have
been if Massengill had volunteered the information. And the
company was under no legal obligation to do so—even in the
face of what appeared to be a mass poisoning caused by its
drug.

On Monday, October 11, in response to Stevenson's
request the day before, the AMA's Paul Leech wired the
headquarters of the S. E. Massengill Company: "KINDLY
TELEGRAPH COLLECT COMPLETE STATEMENT OF COMPOSI-

TION ELIXIR SULFANILAMIDE MASSENGILL." At the time that this telegram was sent, Ivo Nelson, a Tulsa pathologist, had completed the autopsies on four of the six elixir victims. He was now certain that the children had been poisoned.

On Tuesday, October 12, Massengill wired its response to the AMA: "ELIXIR SULFANILAMIDE CONTAINS SULFANIL-AMIDE 40 GRAINS TO EACH FLUID OUNCE DISSOLVED IN A MIXTURE OF TWENTY FIVE PERCENT WATER AND SEVENTY FIVE PERCENT DIETHYLENE GLYCOL WITH MINUTE QUANTITIES OF FLAVOR AND COLOR." Massengill executives asked the AMA to keep the specific contents of their product confidential; they reassured the AMA that the drug had been tested and was safe. The executives also argued that if, as the AMA feared, Elixir Sulfanilamide was poisonous, they would have expected many more people to have been affected—a prediction that would soon come true. The AMA didn't trust Massengill's description of the ingredients in its elixir. To confirm that the drug contained what Massengill had claimed—and that it didn't contain any undeclared heavy metals such as lead, arsenic, or mercury—the AMA's chemical laboratory analyzed it. The scientists found that it contained 10 percent sulfanilamide, 15 percent water, and trace amounts of raspberry extract, caramel, and saccharin, for coloring and flavor. But the main ingredient—and the one that distinguished it from every other sulfanilamide product on the market—was 72 percent diethylene glycol, the solvent that had allowed the antibiotic to go into solution so easily. When asked, Massengill had accurately reported the contents of its product to the AMA. What Massengill hadn't realized was just how dangerous one of those ingredients could be.

On Monday, October 18, the FDA—although it had little to no authority in the matter—sent its chief medical officer, Ted Klumpp, and its chief inspector, Bill Ford, to Bristol to visit Massengill. At the time, the S. E. Massengill Company had two hundred employees in Tennessee, a branch plant in the Midwest, sales offices in New York and San Francisco, and nearly two hundred salesmen spread across the country. Prior to the visit, Massengill's lawyer, who was also the owner's son-in-law, had claimed that the company had done thorough safety testing before releasing 240 gallons of Elixir Sulfanilamide. When the FDA officials visited Massengill, however, they found that the company hadn't performed any safety tests and that its chemists weren't conducting any new testing.

Klumpp and Ford then met Massengill's chief chemist, Harold Watkins, who admitted that his product had never been formally tested for safety. He did say, however, that after the Tulsa reports he had fed the drug to a few guinea pigs and that all the animals were "perfectly well." In fact, according to Watkins, they seemed to like it. Watkins also said that he had ingested some of the elixir himself, later writing, "On Monday, October 18, I started taking the elixir, one ounce a day, in two doses, and am still here and writing to you, though I have taken four ounces in four days. Federal men and our men saw me take it." If anything, he argued, he had urinated more than normal afterward (actually an early sign of kidney disease). Klumpp found Watkins uncomfortably glib and arrogant. "Instead of properly testing this article prior to its distribution," wrote Klumpp, "the chemist attempts to atone for it by a display of heroics." Klumpp later denied ever having seen Watkins take the drug.

During the investigation, Massengill executives continued to argue that the product wasn't the problem, stating, "The drug sulfanilamide had been so exploited by physicians and the press that everyone in the country was going wild with it and using it for everything and now the disastrous effects were coming out." In other words, the executives argued, children were dying from side effects caused by overdosage and overuse of sulfanilamide, not from any specific ingredient in the product. They also argued that the problem might have been caused by the mixture of their elixir with other drugs. Klumpp, sickened by Massengill's attitude toward drug development, wrote, "The only criteria of [Elixir Sulfanilamide's] safety and value as a medicine were that [when] the ingredients were mixed, [they] did not immediately explode."

A closer look at Massengill's chief chemist was far from reassuring. Watkins, it appeared, had led a Dickensian, itinerant, and, in many ways, dishonest life. Harold Cole Watkins was born on January 25, 1880, in Paris, Maine, the first child of George Watkins, the publisher of a small newspaper, and Annah Cole, the youngest daughter of a local judge. When Harold was three years old, his mother died of tuberculosis. The following year, his youngest sister also died of tuberculosis. When Harold was five, his father remarried and moved to Portland, Maine; George's new wife died less than two years after the wedding. Harold had watched his mother, stepmother, and younger sister die before he was six years old. When he was seven, Harold was sent to live with a family member whom he had never met. When he was ten, Harold's father also died.

Before taking the position at Massengill, Harold Watkins had trouble holding a job. After receiving a degree in pharmacy from the University of Michigan in 1901, Watkins worked in Omaha, Nebraska, from 1903 to 1906; Portland, Maine, in 1906; New York City from 1907 to 1908; Anacortes, Washington, from 1908 to 1910; Spokane, Washington, from 1910 to 1911; Vancouver, Washington, from 1911 to 1912; Portland, Oregon, in 1912; Ogden, Utah, from 1912 to 1913;

Sacramento, California, from 1913 to 1917; and Brooklyn, New York, from 1917 to 1920. In 1917, while working for a wholesale firm in Sacramento, Watkins pleaded guilty to stealing a silver dish from his employer and later selling it to a local assayer for $85. With a wife and one small child, Watkins avoided jail time by throwing himself on the mercy of the court. He was placed in the custody of the county sheriff until he could come up with the $3,000 bail.

In 1929, Watkins launched a company that sold a weight-loss product called Takoff, which he claimed produced "perfect slenderness" and a "trim, youthful, athletic look." To avoid charges of mail fraud, Watkins later agreed to remove Takoff from the market.

This was the man whom Massengill had hired to make a product that would soon be ingested by hundreds of Americans, including many children.

———

By October 18, 1937, when FDA officials visited Massengill, the company had sent out more than one thousand telegrams to pharmacists and physicians: "DO NOT USE ELIXIR SULFANILAMIDE SHIPPED. RETURN OUR EXPENSE." But the telegram had failed to mention the possible fatal outcomes. The FDA insisted that a second telegram be sent, this time explaining how dire the situation had become. Massengill complied: "IMPERATIVE YOU TAKE UP IMMEDIATELY ALL ELIXIR SULFANILAMIDE DISPENSED. PRODUCT MAY BE DANGEROUS TO LIFE. RETURN ALL STOCKS, OUR EXPENSE."

In November 1937, the FDA was poised to launch the greatest manhunt in the history of the federal agency. Unfortunately, as described in an editorial in the *Journal of the American Medical Association*, the FDA was "as inefficiently armed as a hunter pursuing a tiger with a fly swatter."

At the time of the Elixir Sulfanilamide disaster, the only federal law governing the sale and distribution of pharmaceutical products was the Pure Food and Drug Act of 1906. The law was born of a book written by a socialist author and a magazine article written by a popular skeptic.

In 1884, Harvey Washington Wiley became the chief chemist at the US Department of Agriculture (USDA). At the time, the food and drug industries were unregulated. Wiley watched helplessly as Americans consumed spoiled meat, sawdust-adulterated flour, and formaldehyde-preserved milk. He insisted that it was time for the federal government to step in. Help came from an unlikely source: Upton Sinclair, a journalist who often railed against the abuses of capitalism. On November 4, 1905, Sinclair published *The Jungle*, a fictional account of the plight of immigrant workers in Chicago's meatpacking industry. Sinclair had intended to hit Americans in their hearts; instead, he hit them in their stomachs. "There would be meat that had tumbled on the floor, in the dirt and sawdust, where the workers had tramped and spit uncounted billions of consumption [tuberculosis] germs," wrote Sinclair.

> There would be meat stored in great piles in rooms; and the water from leaky roofs would drip over it, and thousands of rats would race about on it. It was too dark in these storage places to see well, but a man could run his hand over these piles of meat and sweep off handfuls of the dried dung of rats. These rats were nuisances, and the packers would put poisoned bread out for them; they would die and then rats, bread and meat would go into the hoppers together.

Sales of meat dropped by half. Following publication of *The Jungle*, Theodore Roosevelt ordered Congress to create legislation guaranteeing pure food. In the end, because of a series of articles in a popular magazine, Roosevelt's new law wouldn't be limited to food.

One month before Sinclair published *The Jungle*, on October 7, 1905, Samuel Hopkins Adams had published the first of a series of articles in *Collier's* magazine, titled "The Great American Fraud." Adams wanted Americans to understand exactly what was contained in patent medicines. So, he sent samples to chemists, finding that many contained alcohol, chloroform, opium, morphine, hashish, and cocaine. Some of these drugs were given to young children. More than five hundred thousand Americans read "The Great American Fraud."

With the public up in arms following the publication of *The Jungle* and "The Great American Fraud," Harvey Washington Wiley felt the time was right to propose a federal law to "cover every kind of medicine for external and internal use," which would require manufacturers to list all ingredients and prohibit them from selling narcotics without a prescription. The bill that President Theodore Roosevelt signed into law, the Pure Food and Drug Act of 1906, was a watered-down version of what Wiley had wanted. If a patent medicine contained alcohol, cocaine, opium, chloroform, or other potentially harmful drugs, manufacturers had to print it on the label. But that was it. They didn't have to prove that their products were safe or effective, and they didn't have to prove that they were made using good manufacturing practices.

Enforcement of the Pure Food and Drug Act fell to the USDA's Bureau of Chemistry. In 1927, the newly minted Food, Drug, and Insecticide Administration took over; three years later, it changed its name to the Food and Drug Administration. Unfortunately, the 1906 law didn't give the FDA much power. Technically, the agency couldn't ask for the withdrawal of a drug just because it was unsafe or because it had been manufactured poorly. It could only ask for withdrawal of a harmful drug if it had been "misbranded or adulterated," or if it had failed to list the presence of alcohol, cocaine, heroin, or cannabis on the label. Infractions were misdemeanors, each necessitating only a small fine. Fortunately, Massengill's Elixir Sulfanilamide was misbranded. By labeling it an elixir, Massengill had implied that it contained alcohol, even though it didn't. Had the drug been called "Solution Sulfanilamide," the FDA would have been powerless to force Massengill to withdraw it, even in the midst of what appeared to be a massive, nationwide poisoning.

On Tuesday, October 19, the day after FDA investigators visited Massengill, the agency set about confiscating every unconsumed drop of Elixir Sulfanilamide. It wouldn't be easy. Some patients in the Deep South were beyond the reach of radio bulletins, and some pharmacists were selling the drug without a prescription, leaving no record of the sale. Also, federal law didn't allow the FDA to compel Massengill to show its shipping records. Working within these restrictions, FDA chief Walter Campbell ordered more than two hundred inspectors and chemists to spread out across the country to investigate every distributor, wholesale druggist, retail druggist, and physician who might have received or prescribed the drug. At one distribution site, more than twenty thousand shipping requests were examined. Not everyone was cooperative. In Texas, one distributor provided the needed information only after he was jailed by authorities.

On Wednesday, October 20, in the middle of the investigation, Samuel Massengill, the CEO of the S. E. Massengill Company, sent a telegram to the AMA: "PLEASE WIRE COLLECT BY WESTERN UNION SUGGESTION FOR ANTIDOTE AND TREATMENT FOLLOWING ELIXIR SULFANILAMIDE." Unfortunately,

there was no antidote. The AMA's response was blunt: "Antidote for Elixir Sulfanilamide-Massengill not known.
Treatment presumably symptomatic."

Elixir Sulfanilamide had been distributed to thirty-one
states ranging from Georgia, which received twenty-one gallons of the drug, to Connecticut, which received only a fraction of a gallon. Massengill shipped six hundred large bottles
of its elixir to pharmacies and distributors and another seven
hundred small sample bottles to physicians' offices and to
sales representatives. Deaths from the drug mounted. By
October 25, forty-six people had died; by October 27, fifty;
by October 31, fifty-eight; by November 2, sixty-one.

When the dust settled, of the 240 gallons shipped, six
were consumed by the public. During the four weeks that the
drug was on the market, 353 people drank it and 105 died;
34 were children. If all 240 gallons of the elixir had been consumed, four thousand people would have been killed.

Physicians were devastated. Perhaps no one more so
than Archie Calhoun of Mount Olive, Mississippi, who had
watched six of his patients die from a drug he had prescribed.
On October 22, 1937, Calhoun wrote an anguished letter
to a local newspaper: "Any doctor who has practiced more
than a quarter of a century has seen his share of death. But
to realize that six human beings, all of them my patients,
one of them my best friend, are dead because they took a
medicine that I prescribed for them innocently . . . well, that
realization has given me such days and nights of mental and
spiritual agony as I did not believe a human could undergo
and survive. I have known hours when death for me would
have been a welcome relief from this agony." Physicians had
prescribed the elixir in one hundred of the 105 deaths.

On Tuesday, November 7, researchers at the University of Chicago solved the mystery of Elixir Sulfanilamide.

In a study published in the *Journal of the American Medical Association*, they injected rats, rabbits, and dogs with either pure sulfanilamide or pure diethylene glycol. Only the animals given diethylene glycol died. Similar to people, they died from kidney failure.

In response to what had been called "one of the most consequential mass poisonings of the twentieth century," Sam Massengill issued a statement. "My chemists and I deeply regret the fatal results, but there was no error in the manufacture of the product. We have been supplying a legitimate professional demand and not once could have foreseen the unlooked-for results. I do not feel that there was any responsibility on our part." Massengill was right to claim that the production and distribution of his drug hadn't violated any laws. At the time of the disaster, companies didn't have to test drugs for safety, nor did they have to list ingredients on the label. But was Sam Massengill right to claim that he couldn't have foreseen the results?

In 1931, six years before the Elixir Sulfanilamide disaster, researchers from the department of pharmacology at Western Reserve University in Cleveland published a paper showing that when diethylene glycol was injected into rats, it was "apt to produce more or less marked pathological changes in the kidney."

In January 1937, eight months before the S. E. Massengill Company shipped Elixir Sulfanilamide across the country, researchers from the department of pharmacology at the Medical College of Virginia in Richmond studying rats and rabbits published a paper showing that "the ingestion of diethylene glycol in concentrations of 3 and 10 percent proved rapidly fatal." The concentration of diethylene glycol in Elixir Sulfanilamide was 72 percent. In both of these studies, investigators found that diethylene glycol caused a dramatic increase in oxalic acid in the kidneys, the direct cause of the kidney failure. (Diethylene glycol is similar to antifreeze, which includes the chemicals ethylene glycol and propylene glycol.)

Harold Watkins was apparently ignorant of these studies. Others weren't. On October 12, 1937, in response to the six deaths of children in Tulsa, Dr. Paul Leech of the AMA's Council on Pharmacy and Chemistry sent a telegram to a doctor in the city. Leech was horrified that diethylene glycol had been used to solubilize sulfanilamide. "PRODUCT CONTAINS MUCH DIETHYLENE GLYCOL AS SOLVENT THAT IS TOXIC AND REPORTED TO CAUSE NEPHROSIS. POSSIBLY MAY BE OXIDIZED TO OXALIC ACID." (Nephrosis is the medical term for kidney disease.) Leech then referred the Tulsa doctor to the 1931 study showing that diethylene glycol caused kidney failure in rats.

In short, the Elixir Sulfanilamide disaster was entirely preventable.

———

On November 30, 1937, FDA chief Walter Campbell wrote a letter to Maise Nidiffer, the mother of little Joan Marlar. Campbell was responding to the letter that Nidiffer had sent to President Franklin Delano Roosevelt three weeks earlier:

> My dear Mrs. Nidiffer,
>
> Believe me when I say, Mrs. Nidiffer, that in the course of my thirty years in the Government service I have never had a more difficult assignment than the President's request to reply to you. Your letter was laid before me during the weeks that we were desperately following up outstanding lots of the Elixir. I have looked at the lovely smiling face of

your little girl and I think I realize in some measure your unending sorrow at a needless sacrifice.

I agree with you that it should not have been necessary for your baby to die to arouse public sentiment to the point that proper legislation will be enacted. If it is enacted—and I sincerely believe that there are strong and good men in both the House and Senate who have resolved that laws will be passed which will make a repetition of this tragedy impossible—it may be of some small comfort to you to know that your letter has had a real influence in bringing about this result.

With the most-sincere sympathy, I am

Sincerely yours,

W. G. Campbell, Chief

On March 5, 1938, as a direct consequence of the Elixir Sulfanilamide disaster, the House of Representatives passed the Food, Drug, and Cosmetics Act; on June 25, President Roosevelt signed it into law. Pharmaceutical companies now had to list all ingredients on the label and, most importantly, they had to perform adequate safety testing before the product would be licensed by the FDA.

━━

Because of weak federal laws, the S. E. Massengill Company was never held accountable for the tragedy of its making. Under the guidelines of the 1906 Pure Food and Drug Act, the government could only fine the company $26,000 for 174 counts of misbranding. In addition, Massengill had to pay $148,000 settling damage claims. Three months later, Sam

Massengill, as the town's largest employer, was elected president of the Bristol Chamber of Commerce.

Following the Elixir Sulfanilamide tragedy, the company thrived. In 1971, it was sold to the London-based pharmaceutical company Beecham, which paid $54 million to acquire two million shares; at the time, about 90 percent of the company's stock was owned by the family. Beecham was attracted by Massengill's popular line of over-the-counter feminine hygiene products. Barbara Martin, in her book *Elixir: The American Tragedy of a Deadly Drug*, summarizes the ironic fate of Sam Massengill and his company: "For a man who conspicuously valued family heritage and civic reputation, he left his namesakes with a vaginal douche and the horror of mass death." Sam Massengill died in 1946 a multi-, multi-millionaire.

Harold Watkins, the company's chief chemist, wasn't as lucky. On January 17, 1939, one week after Sam Massengill had been elected president of the Bristol Chamber of Commerce, Harold Watkins was found dead by his wife in the kitchen of their small brick bungalow: lying facedown, his spectacles lodged under his body, and a .38 caliber automatic pistol a foot from his head.

CHAPTER 6

ANNE GOTTSDANKER

Vaccines

IN MANY WAYS, ANNE GOTTSDANKER'S LIFE WAS IDYLLIC.

Her parents, both of whom were psychologists, moved to Santa Barbara, California, in 1949, two years before she was born. "We lived in an older part of town," Gottsdanker recalled. "It was a big two-story house with wooden floors and wooden frames with a huge sun room within walking distance of Santa Barbara State Teachers College [now University of California, Santa Barbara]." Anne shared a room with her brother, Jerry, who was five years older.

Gottsdanker remembers "sparkling conversations around the dinner table about art, politics and travel." She remembers her parents doing the dishes together. She remembers camping trips to Yosemite. "I was close with my father," she said. "He would make up stories at bedtime. He wasn't demonstrative, but he really cared about me. He came from a bigger family and was more relaxed than my mother, Josephine, who was an only child."

On Monday afternoon, April 18, 1955, Josephine Gottsdanker drove Anne and Jerry to the pediatrician. Several days earlier, she had watched the television program *See It Now*. Edward R. Murrow, a CBS news correspondent, was interviewing Jonas Salk, the scientist who had just invented a polio vaccine.

Summer was near and, like most American mothers in the 1950s, Josephine was afraid. Afraid of other children. Afraid of swimming pools, water fountains, city streets, recreational camps, church groups, sporting events, and public gatherings. Afraid that this would be the summer that her children would be among the tens of thousands claimed every year by polio.

Anne Gottsdanker still remembers that doctor visit. She remembers the nurse taking a vial of fluid out of the refrigerator, drawing it into a sterilized glass syringe, and injecting it into the muscle of her right leg. Minutes later, the procedure was repeated for her brother.

On April 22, four days after the injections, Josephine Gottsdanker loaded her children into the back seat of their car for the 320-mile, five-hour trip to Calexico, California—a town on the border of California and Mexico—to visit relatives.

The visit was uneventful. But on the afternoon of April 26, driving back from Calexico—and eight days after receiving the Salk polio vaccine—Josephine noticed that something was wrong with her daughter: "We stopped at a little mountain village for coffee and ice cream and she said that her head hurt. I thought that her ponytail was pulled too tight. It seemed to me like a casual child's complaint at the time. Then she vomited in the car. We took her to County Hospital. By then she had lost motion in the upper part of her leg—then it moved to the lower part."

"I remember throwing up," said Anne. "I remember it was like the flu." Mostly, however, she remembers her helplessness as the mysterious disease progressed. "I remember my dad taking me out of the car and carrying me to the hospital. I remember lying in the hospital and not being able to move anything. I was totally paralyzed. I didn't know what was going on, but I was too young to be afraid."

Anne's right leg was completely paralyzed. The diagnosis: polio. Two features of her polio, however, were unusual. First, it was April; polio was a summer disease. Second, she was paralyzed in the same leg where she had been vaccinated. Could the polio vaccine actually be causing polio?

Over the next few months, the answer would become clear. Salk's polio vaccine had caused a man-made polio epidemic. Tens of thousands of children would be temporarily paralyzed, hundreds would be permanently paralyzed, and ten would be killed. Federal regulators immediately suspended the polio-vaccine program until they could figure out what had gone so horribly wrong. And Jonas Salk would be forever shunned by his colleagues.

What happened? And, more importantly, why don't we know this story?

———

Salk wasn't the first scientist to make a polio vaccine. Between 1934 and 1935, John Kolmer of Philadelphia and Maurice Brodie of New York had also tried.

Kolmer injected monkeys with poliovirus, removed their spinal cords, ground them up, suspended them in a salt solution, filtered them through a fine mesh, and treated them with ricinoleate for two weeks. (Ricinoleate, a chemical derived from the castor plant, was also used to make soap.) Kolmer assumed that the ricinoleate would kill poliovirus so that it could no longer cause disease but would still induce immunity. He injected his vaccine into his eleven- and fifteen-year-old sons; his assistant, Anna Rule; and twenty-five children. Declaring his vaccine safe, Kolmer then injected ten thousand more children.

Maurice Brodie's vaccine was similar to John Kolmer's. The only difference was that instead of killing poliovirus with ricinoleate, Brodie used formaldehyde, which was popular among morticians to preserve dead bodies. Brodie inoculated himself, five coworkers, and twelve children. Believing that he, too, had made a safe polio vaccine, Brodie then inoculated seven thousand more children.

In 1935, Kolmer and Brodie presented their findings to hundreds of physicians, scientists, and public health officials at a meeting of the American Public Health Association in Saint Louis. Kolmer spoke first. He said that three doses of his vaccine had been given to ten thousand children in thirty-six states and Canada. Unfortunately, Kolmer didn't include unvaccinated children in his study, so he couldn't tell whether his vaccine really worked. Worse, several children developed polio soon after receiving the vaccine:

- Sally Gittenberg, a five-year-old from Newark, New Jersey, was injected with Kolmer's vaccine in her left arm. Twelve days later, she developed fever, a headache, a stiff neck, and vomiting. Sixteen days later, her left arm was completely paralyzed.

- Esther Pfaff, a twenty-one-month-old from Westfield, New Jersey, was injected with Kolmer's vaccine in her right arm. Two weeks later, Esther couldn't move the arm. Three weeks later, she was dead.

- David Costuma, an eight-year-old from Plainfield, New Jersey, was injected with Kolmer's vaccine in his right arm. Two weeks later, suffering from headaches, tremors, and paralysis of his right arm, David was taken to the hospital. The following day, he was dead.

- Hugh McDonnell, a five-year-old also from Plainfield, New Jersey, was injected with Kolmer's vaccine in his left arm. One month later, he developed paralysis of the same arm. Two days later, he was dead.

In the end, Kolmer's vaccine paralyzed ten children and killed five. Most were paralyzed within a few weeks of receiving their vaccine and all were paralyzed in the arm that was inoculated. Sally Gittenberg was the only child in Newark to develop polio that month, and the two children in Plainfield were the only ones to die from polio in the city. Kolmer argued that these children had been infected with natural poliovirus, not live poliovirus inadvertently contained in his vaccine. "I do not personally believe that the vaccine was responsible for these cases," he said.

Angered by Kolmer's denials, James Leake, medical director of the Public Health Service in Washington, DC, rose to speak. "Jimmy Leake then point-blank accused Kolmer of being a murderer," recalled one observer. "[He] used the strongest language that I have ever heard at a scientific meeting and when he got through speaking, both vaccines were dead. When you say someone is committing murder, people usually stop and think." Kolmer was thoroughly humiliated: "Gentlemen, this is one time I wish the floor would open up and swallow me."

Unlike Kolmer, Maurice Brodie included both vaccinated and unvaccinated children in his study. He found that five of the 4,500 unvaccinated children contracted polio, but only one of the seven thousand vaccinated children got the disease. The vaccine, apparently, worked. But Brodie's description of one case was particularly troublesome: a twenty-year-old man who developed paralysis in the inoculated arm and died four days later. Several months after the meeting, James Leake, in an article published in a medical journal, further questioned the safety of Brodie's vaccine when two more children—ages five months and fifteen months—developed polio within two weeks of receiving it.

The men involved in the first polio vaccines met dramatically different fates. John Kolmer continued to publish scientific papers, rose to the level of professor at Temple University School of Medicine, and retired in 1957. Maurice Brodie, on the other hand, was fired from his jobs at New York University and the New York City Health Department. In May 1939, when he was only thirty-six years old, Brodie died. Many at the time and since have speculated that he killed himself.

The vaccine trials of John Kolmer and Maurice Brodie had a chilling effect on polio vaccines. Twenty years passed before anyone dared to try again.

━━

Jonas Salk was born on October 28, 1914, in a tenement in East Harlem, New York—the first son of Russian immigrants and the eldest of three brothers. After finishing four years of high school in three, Salk entered the City College of New York and later won a scholarship to the medical school at New York University. In December 1941, after the United States entered World War II, Salk was given a choice. He could either be commissioned as a doctor in the armed forces or remain in the United States to pursue a scientific career. He chose science, working on an influenza vaccine in the laboratory of Dr. Thomas Francis at the University of Michigan. Fifteen years later, Francis would supervise the critical test of Salk's vaccine.

In 1943, while Salk was working on the influenza vaccine, ten thousand people, mostly children, contracted polio in the United States; in 1948, when Salk was first studying polioviruses at the University of Pittsburgh, twenty-seven thousand more people were affected; and in 1952, when Salk was first testing his polio vaccine in and around Pittsburgh, fifty-nine thousand more cases occurred. A national poll found that polio was second only to the atomic bomb as the thing Americans feared most.

There was a desperate, growing desire to prevent polio.

———

Viruses, unlike bacteria, grow inside cells. To grow polioviruses, John Kolmer and Maurice Brodie had used cells from monkey brains and monkey spinal cords. Salk, on the other hand, used cells from monkey testicles. Later, concerned that people would never accept a vaccine grown in monkey testicles, Salk switched to monkey kidney cells, which are still used to make polio vaccines today.

Because three different types of poliovirus cause disease, Salk knew that he would need to include representatives of all three types in his vaccine. For type 1, Salk chose the Mahoney strain—a decision that would haunt him for the rest of his life. The Mahoney strain was first recovered from a child in Akron, Ohio, whose last name was Mahoney. But the Mahoney strain wasn't limited to the Mahoneys. The Klines, living next door, were also infected. Three of five Kline children were paralyzed and later died from the disease—an early clue to the unique horror of this particular strain. The other two strains of virus contained in Salk's vaccine, representing types 2 and 3, weren't controversial.

The Mahoney strain had one more characteristic that made it particularly dangerous. Although all strains of poliovirus paralyzed monkeys after injection into their brain or spinal cord, only a few strains paralyzed monkeys after injection into their muscles—the route by which children would soon be inoculated. After intramuscular injection, the Mahoney strain was ten thousand times more likely to travel to the brain and spinal cord than any other polio strain. By choosing Mahoney, Jonas Salk was about to jump without a net.

To make his vaccine, Salk took the three strains of poliovirus and, like Maurice Brodie before him, treated them with formaldehyde. To determine how long it took to kill these three viruses, Salk periodically injected them into the brains of monkeys and waited to see if the monkeys would become paralyzed. He found that the viruses no longer paralyzed monkeys after about ten days of formaldehyde treatment. Then, as a safety measure, Salk treated the viruses for another two days. The strain of virus that took the longest to kill was Mahoney.

———

To determine whether his vaccine worked and was safe, Salk went to the D. T. Watson Home for Crippled Children. After injecting children with what he hoped would be a safe and effective vaccine, Salk couldn't sleep. "He came back again that first night to make sure that everyone was all right," recalled the nurse superintendent. "Everyone was." The results were disappointing. Salk found that only one of his three vaccine strains induced high levels of protective antibodies in the blood.

In May 1952, Salk tried again. This time, he injected children at the Polk State School, which housed "mentally retarded" boys and men. Unlike the children at D. T. Watson, however, the Polk residents developed immunity to all three types of poliovirus. Also, the vaccine appeared to be safe. "I've got it," Salk said to his wife.

In the spring of 1953, Salk performed the time-honored ritual of vaccinating himself, his wife (Donna), and his three children (Peter, age nine; Darrell, age six; and Jonathan, age three) with his vaccine. "It is courage based on confidence, not daring," he said. "Our kids were lined up to get the vaccine," recalled Donna Salk. "I had complete and utter confidence in Jonas."

To ensure that his vaccine was safe—and that it couldn't possibly contain live poliovirus—Salk relied on a mathematical model that would soon become controversial. He called it the straight-line theory of inactivation, and it would be at the center of the tragedy that would soon follow. The straight-line theory defined exactly how long it took to completely kill every single particle of live poliovirus. Salk was certain that he could avoid the tragic mistakes made by Kolmer and Brodie.

Here's how Salk's straight-line theory of inactivation worked: One dose of the vaccine given to children at the D. T. Watson and Polk schools was about one-fifth of a teaspoon. Salk showed that the quantity of live poliovirus contained in one dose before treatment with formaldehyde was about one million infectious virus particles. Treatment with formaldehyde caused the quantity of live virus to decrease steadily, predictably, and exponentially. After treatment for twelve hours, the quantity of live virus was one hundred thousand infectious particles; after treatment for twenty-four hours, ten thousand infectious particles; and after treatment for seventy-two hours, only one infectious particle remained—the other 999,999 had been killed by formaldehyde. In other words, the amount of live virus contained in the vaccine had been reduced one-million-fold in three days.

When Salk plotted the quantity of live virus against the length of time that virus was treated with formaldehyde, the points connected in a straight line. Salk reasoned that if the virus was treated for three more days, and the line remained straight, then another million-fold reduction would occur. Now, instead of having one infectious particle in one dose, there would be one infectious particle in one million doses. Treatment for another three days would then reduce the quantity of live virus to one infectious particle in one trillion doses—more vaccine than would be required to vaccinate everyone living on the planet. For all practical purposes, treatment for nine days would completely inactivate poliovirus. All this, of course, assumed that the line remained straight.

The graph constructed by Salk consisted of two parts: the part a researcher could measure in a laboratory and the part that Salk *assumed* to be true. It was possible to test preparations of virus to find that they contained one million or ten thousand or one thousand or one hundred or ten or even one infectious particle per dose, but it was not practical to test a million or a trillion doses to make sure that no live virus remained. For several years after the events in the spring of 1955, Salk was the only scientist in the world who believed that his straight-line virus-inactivation curve remained straight.

—

In 1954, the National Foundation for Infantile Paralysis (otherwise known as the March of Dimes) footed the bill for what would become the largest test of a medical product in history. About twenty thousand physicians and health officers, forty thousand registered nurses, fourteen thousand school principals, fifty thousand teachers, and two hundred thousand Americans in forty-four states volunteered. The vaccine was made by two veteran vaccine makers: Eli Lilly and Parke Davis. Four hundred and twenty thousand children were injected with Salk's vaccine, two hundred thousand were injected with saltwater, and 1.2 million were given nothing and merely observed—a total of 1.8 million participants. The choice to include children inoculated with saltwater and children who weren't inoculated with anything enabled researchers to determine whether the vaccine actually worked. On Monday, April 26, 1954, at 9 a.m., six-year-old Randy Kerr of McLean, Virginia, received the first shot. Because the vaccine and placebo injections were coded, neither Randy, his parents, nor the nurse administering the shot knew what he was getting. Randy smiled for the cameras.

During the next five weeks, a series of three shots of vaccine or saltwater was given to first, second, and third graders across the country. Shots were administered in the left arm. Children who participated in the study received a lollipop, a pin that read "Polio Pioneer," and a chance to avoid the crippling effects of polio. (A trial of this size today would cost about $6 billion.)

The results were encouraging:

- Sixteen children in the study died from polio—all had received the placebo. (This statistic has always haunted me. Had these children not had the misfortune to have been randomized to the placebo group, they could possibly still be alive and well today.)

- Thirty-six children were permanently paralyzed or placed in iron lungs to help them breathe. Thirty-four of the thirty-six received placebo. (A statistic that is equally haunting.)

The vaccine worked. But was it safe? Although several cases of polio had occurred within two months of receiving the vaccine, Thomas Francis, the director of the trial, concluded that Salk's vaccine wasn't the cause. And he was right—it wasn't. None of the children were paralyzed in the arm that was inoculated, clearly different from those who had been paralyzed by the Kolmer and Brodie vaccines. About 420,000 children had been given Salk's vaccine and not one had been paralyzed by it. The vaccine, apparently, was safe. Now it could be given with confidence.

On April 12 at 10:20 a.m.—ten years to the day after the death of polio's most famous victim, Franklin Delano Roosevelt—Thomas Francis stepped to the podium at Rackham Hall on the campus of the University of Michigan to announce the results of the groundbreaking March of Dimes study. More than five hundred people—including 150 press, radio, and television reporters—filled the room; sixteen television and newsreel cameras stood on a long platform at the back; and fifty-four thousand physicians in movie theaters across the country watched the broadcast on closed-circuit television. Americans turned on their radios, department stores set up loudspeakers, and judges suspended trials so that everyone in the courtroom could hear what Francis was about to say. Europeans tuned in to Voice of America.

Before Francis presented the results, copies of the report were distributed to the press. "They brought the report in on dollies, and newsmen were jumping over each other and screaming, 'It works! It works! It works!'" one reporter remembered. "The whole place was bedlam. One of the doctors [had] tears in his eyes."

When Francis finished explaining that Jonas Salk's polio vaccine worked and was safe, church bells rang across the country, factories observed moments of silence, synagogues and churches held special prayer meetings, and parents and teachers wept. One shopkeeper hung a banner on his window that read, "Thank you, Dr. Salk." Newspapers across the country bore the same headline, "SAFE, POTENT, AND EFFECTIVE." "It was as if a war had ended," one observer recalled.

But the war hadn't ended.

On Sunday, April 24, 1955, J. E. Wyatt, the health officer in charge of Idaho's southeast district, received a phone call from a local physician about a girl in Pocatello. "I've just seen a youngster who seems to have polio," the doctor said. "Her mother says she noticed a little stiffness of the neck yesterday and she had fever. Today her left arm became paralyzed. Her name is Susan Pierce." Wyatt, reassured by the results of the large field trial, was certain that Salk's vaccine was safe. "She must have been exposed [to natural poliovirus] before the vaccination," he told the physician, "and there wasn't time enough for the vaccine to protect her. But I'm glad you called. We'll keep a close watch on things."

Susan Pierce had been injected in her left arm with Salk's vaccine on April 18, 1955. Five days later, she developed fever and a stiff neck. Six days later, her left arm was paralyzed. Seven days later, she was placed in an iron lung. Nine days later, she was dead.

Other cases in Idaho soon followed: On April 26, six-year-old Jimmy Shipley was admitted to the hospital with paralysis of his left arm. On April 30, Bonnie Gale Pound of Lewiston was admitted to St. Joseph's hospital. The Idaho Air National Guard flew an iron lung in from Boise in an unsuccessful attempt to save her life. Between May 1 and 3, seven-year-old Jimmy Gilbert of Orofino developed paralysis in his left arm, eight-year-old Dorothy Crowley of Ahsahka was placed in an iron lung, and Janet Lee Kincaid of Moscow and Danny Eggers of Idaho Falls died from polio.

These children shared one thing in common: they had all been paralyzed in the arm that was inoculated. Also, and perhaps equally worrisome, all had suffered or died from polio in April and early May. Although hundreds of children in Idaho were paralyzed by polio every year, the first few cases didn't usually appear until June.

Salk's vaccine was made by five different pharmaceutical companies: Eli Lilly and Parke Davis (both of which had made the vaccine for the large field trial) and three smaller companies, Pitman-Moore, Wyeth, and Cutter Laboratories. The Idaho schoolchildren, however, had received a vaccine made by only one company: Cutter Laboratories of Berkeley, California.

On April 28, 1955, the day that Cutter recalled its vaccine, the surgeon general of the United States, Leonard Scheele, called on Alexander Langmuir, the chief epidemiologist at the Communicable Diseases Center in Atlanta, Georgia, to investigate the tragedy. (The Communicable Diseases Center would later become the CDC). Five years earlier, Langmuir had created what he called the Epidemic Intelligence Service to respond to biological attacks from enemies outside the United States. He never imagined that his first assignment would be to investigate a biological attack from inside the country.

Langmuir's team found that two specific lots of Cutter's vaccine had paralyzed fifty-one children and killed five. They also found that children given Cutter's vaccine were more likely to be paralyzed in their arms, more likely to suffer severe and permanent paralysis, more likely to require iron lungs, and more likely to die than children naturally infected with polio. The reason that Cutter's vaccine was more deadly was that the company had failed to adequately kill the most lethal of all the poliovirus strains: Mahoney.

An orthopedic surgeon in Idaho, Manley Shaw, soon found that Langmuir's team had discovered only the tip of the iceberg. Shaw examined the medical records of more than four hundred school children from Pocatello, Boise, and Lewiston and found that one of every three children injected with the two contaminated lots of Cutter's vaccine developed fever, sore throat, headache, vomiting, muscle pain, stiff neck, stiff back, and a slight limp. Many of these children had persistent muscle weakness months later. Because the two contaminated lots contained 120,000 doses, this meant that forty thousand vaccinated children had developed symptoms of polio.

The Cutter tragedy wasn't limited to those who were injected. On May 8, eleven days after the recall, a twenty-eight-year-old mother visiting Atlanta from Knoxville was hospitalized with polio and put on a ventilator. She had never received Cutter's polio vaccine, but her children had. On May 9, another young mother, who was a neighbor of Alexander Langmuir's secretary in Atlanta, died of polio. Like the Knoxville woman, the neighbor had never received a polio vaccine, but her baby had received Cutter's vaccine one month earlier.

All of this meant that the live, deadly poliovirus in Cutter's vaccine was now spreading through the community. Subsequent investigations found that seventy-four unvaccinated family members had been paralyzed. One was a mother who was eight months pregnant; on May 12, she died of polio after contracting the disease from her one-year-old son.

When the man-made polio epidemic caused by Cutter's vaccine finally ended, seventy thousand people had developed short-lived polio, 164 were severely paralyzed, and ten were killed. It was the worst biological disaster in the history of the United States. "I wish we'd never heard of the vaccine," said Robert Cutter, CEO of Cutter Laboratories. Alexander Langmuir called it "the Cutter Incident."

Salk had shown in his studies in and around Pittsburgh that he could make a safe and effective vaccine. Thomas Francis had shown that Salk's vaccine was safe in the largest trial of a medical product in history. In the hands of Cutter Laboratories, however, Salk's vaccine was now paralyzing and killing children as well as family members and people in the community. What went wrong at Cutter Laboratories?

First, Cutter used the Mahoney strain. Salk had picked that strain because it evoked the best immune response. He knew that Mahoney was deadlier than any other type 1 strain, but he reasoned that if the virus was completely inactivated with formaldehyde, it didn't matter. Nonetheless, Salk's choice of Mahoney left vaccine manufacturers with little room for error. If Cutter had chosen a strain other than Mahoney, the paralysis and death caused by the vaccine probably would never have happened. However, all five companies making a polio vaccine in the spring of 1955 used Mahoney. So, Cutter can't be blamed for its choice.

Second, viruses grow in cells. In order for formaldehyde to completely inactivate poliovirus, all cells and cell debris had to be completely removed by filtration before inactivation. If not, poliovirus could hide within cell debris, untouched by formaldehyde. In other words, filtration was key.

Salk had used a Seitz filter, which was made of thick layers of asbestos. Filtration through Seitz filters was thorough, resulting in liquids those in the beverage industry called "gin pure." Unfortunately, because Seitz filtration was slow—and because Cutter Laboratories was now charged with making hundreds of thousands of doses—researchers at Cutter chose glass filters, which were faster but less efficient. Glass filters were made by partially fusing heated glass so that it contained progressively smaller holes. Fluids containing poliovirus and cell debris would first pass through glass filters with large holes (coarse filters) then through filters with smaller and smaller holes (medium, fine, and ultrafine filters). The goal was to remove every speck of cell debris.

Parke Davis used glass filters and made a vaccine that was safe. So why did Cutter Laboratories have a problem? As it turned out, not all glass filters were the same. Their efficiency depended on the craftsmen who made them. And Cutter had chosen a poorer craftsman. Again, Cutter can't be blamed for using glass filters. But if, like Eli Lilly and Pitman-Moore, Cutter had chosen to use Seitz filters, the Cutter Incident would never have happened.

Third, safety tests were inadequate. To determine whether live poliovirus was inadvertently present in the final product, the federal government required two safety tests: vaccine virus had to be injected into monkey spinal cords and inoculated onto monkey kidney cells in the laboratory. If monkeys weren't paralyzed in thirty days, and if cells in the laboratory didn't die in fourteen days, then the vaccine was considered to be free of live virus. Unfortunately, these safety tests were woefully inadequate. One month after the Cutter Incident, federal regulators changed their requirements so that much larger quantities of vaccine had to be inoculated onto monkey kidney cells for much longer periods of time. Regulators also insisted that monkeys be given a drug to suppress their immune system before they were injected with the vaccine. This made it five hundred times more likely that the monkeys would be paralyzed if small amounts of live poliovirus were present.

Again, Cutter can't be blamed for the lack of adequate safety testing in the spring of 1955. If wasn't until after the Cutter Incident that researchers realized that safety testing was inadequate. But if better safety tests had been available earlier, the Cutter Incident would never have happened.

Fourth, Cutter let filtered virus sit in the refrigerator for weeks before inactivating it with formaldehyde. As a consequence, small clumps of monkey kidney-cell debris collected on the bottom of the flask, preventing the formaldehyde from adequately killing every single live virus particle. Among pharmaceutical companies, only Cutter let filtered virus sit around that long. But again, the lack of accurate, federally mandated safety testing prevented Cutter from knowing it had a problem.

Fifth, and most important, Cutter never constructed a graph to prove that formaldehyde was killing the virus reproducibly and in a straight line. Salk had recommended testing at least four samples during the inactivation process and said that the last sample should be free of live virus. If manufacturers found that it took three days to eliminate live virus, then Salk recommended treatment with formaldehyde for an additional six days. If it took five days, the preparation should be treated for an additional ten days. This recommendation provided the margin of safety that Salk deemed critical to the manufacture of a safe vaccine.

Eli Lilly and Parke Davis both tested six samples during inactivation. Wyeth tested five. Pitman-Moore tested only three samples but always found that it could eliminate live virus after three days of treatment. Cutter Laboratories only tested two samples and *never* detected a sample that was free of live virus. Therefore, the researchers weren't able to determine how long to treat with formaldehyde. No company showed a greater disregard for Jonas Salk and his theories than Cutter Laboratories.

Sixth, officials at Cutter Laboratories didn't tell anyone, including federal regulators, that they were having a problem. And they never consulted the man who was most qualified to help them: Jonas Salk. Although Walter Ward, the person in charge of the polio-vaccine program at Cutter Laboratories, wrote several letters to Salk before Cutter sold its vaccine, he never mentioned that his company was having trouble inactivating poliovirus. In retrospect, a remarkable omission.

Seventh, during the field trial of Salk's vaccine, the March of Dimes, which supervised the trial, required Eli Lilly and Parke Davis to make eleven consecutive lots of vaccine that passed safety testing. After the federal government took over the polio-vaccine program, it dropped this consistency requirement. As a consequence, the government had no way of knowing that Cutter was having a problem. Cutter never made eleven consecutive lots of vaccine that passed safety tests. In fact, it never made four such lots.

Cutter did many things wrong. It let filtered virus sit in the refrigerator before treating it with formaldehyde; it tested the fewest number of samples during inactivation; it never constructed a graph to figure out how long to inactivate virus; it used the smallest quantity of vaccine for safety testing on monkey kidney cells; and it didn't have the internal expertise that was available to other companies. As a consequence, Cutter Laboratories distributed a polio vaccine that was far more dangerous than any other polio vaccine made in the United States or in the world. "They just blindly followed a protocol without thinking about it," recalled a senior scientist at another company. "They didn't have the expertise to think about it."

When Cutter Laboratories pulled its vaccine from the market, public-health officials and federal regulators, believing that they had solved the problem, breathed a sigh of relief. But Alexander Langmuir had been wrong to label the tragedy the Cutter Incident. Because Cutter Laboratories wasn't the only American company to make a vaccine that paralyzed and killed children in the spring of 1955.

———

Pamela Erlichman was seven years old when she was injected with Jonas Salk's polio vaccine at her school in Bucks County, Pennsylvania. Within three days, she couldn't move her left arm, couldn't breathe on her own, and died from polio. Pamela's father, Fulton Erlichman, a local pediatrician, called the school to find out which vaccine his daughter had received. Erlichman assumed that she had been injected with one of the two contaminated lots of Cutter's vaccine. But Pamela hadn't received Cutter's vaccine. She had received a vaccine made by Wyeth—specifically, lot number 236. Langmuir's team found that Wyeth's vaccine paralyzed eleven children, three in the inoculated arm. Also, similar to Cutter's vaccine, children immunized with Wyeth's vaccine appeared to be spreading the virus to family members and community contacts, some of whom were also severely paralyzed.

On April 30, 1955, when it was apparent that Cutter's vaccine was dangerous, industry representatives met to discuss the problem. All of them lamented that many of their lots had failed safety tests and had to be discarded. Although Langmuir had labeled the tragedy in the spring of 1955 the

Cutter Incident, it would have been more accurate to call it the "Scale-Up Incident." All of the companies had trouble inactivating poliovirus for mass production—even Eli Lilly and Parke Davis, which had made the vaccine for the field trial.

===

Although everyone who worked in the federal regulation of the polio vaccine in 1955 was fired, no one suffered from the Cutter tragedy more, or more unfairly, than Jonas Salk.

Salk died on June 23, 1995, five years before the United States considered his vaccine to be the last, best vaccine to prevent polio. On the world's stage, Salk was a respected figure. But among fellow scientists—who were always suspicious of his straight-line theory of inactivation—he wasn't. At least eight polio researchers in the 1930s, 1940s, and 1950s were elected to the National Academy of Sciences, a prestigious society for those considered to have made important contributions. Salk wasn't among them. Renato Dulbecco, winner of the Nobel Prize for his work on viruses that cause cancer, wrote Salk's obituary: "For his work on polio vaccine, Salk received every major recognition available in the world from the public and governments. But he received no recognition from the scientific world—he was not awarded the Nobel Prize, nor did he become a member of the U. S. National Academy of Sciences. The reason is that he did not make any innovative scientific discovery." Until the day he died, Jonas Salk was damned by the Cutter tragedy.

===

The Cutter Incident was the first coordinated national response to a public-health emergency and was a turning point in the history of the CDC. Since its inception, the Epidemic Intelligence Service—dramatized in movies such as *And the*

Band Played On, *Outbreak*, and *Contagion*—has monitored diseases such as anthrax, SARS, Ebola, MERS, and
COVID-19. Epidemic Intelligence Service officers have determined how specific diseases—most notably AIDS, Ebola, and
novel coronaviruses—are spread.

The Cutter tragedy also gave birth to tighter government
regulations. At the time of the polio-vaccine disaster, federal
regulation of vaccines was in its infancy. On July 15, 1955,
only three months after the incident, the government created
the Division of Biologics Standards—a separate department
within the National Institutes of Health. By 1956, the number of professionals regulating vaccines increased from ten to
150, and regulators were now actively involved in studying
the vaccines they supervised. A series of consecutive lots of
vaccine equal in potency, safety, and efficacy continues to be
required of all vaccine makers today, and the term "consistency lots," born of the Cutter Incident, is still used.

Between 1956 and 1961, when more than four hundred
million doses of Salk's polio vaccine were distributed, the incidence of paralysis in the United States declined by 90 percent. Salk's vaccine worked and was safe. Nonetheless, in the
early 1960s, his polio vaccine was replaced by a different vaccine made by Albert Sabin. Sabin made his vaccine by taking
poliovirus and weakening it in the laboratory. This vaccine
was attractive because it was cheaper and easier to administer (a squirt in the mouth instead of a shot in the arm). By the
late 1970s, polio was eliminated from the United States. But
one terrible problem remained: Sabin's vaccine could actually
mutate back to a virus that resembled natural poliovirus. It
was rare—only occurring in one of every 2.4 million doses—
but it was real. As a consequence, in the 1980s and 1990s the
only polio that occurred in the United States was caused by
Sabin's live, weakened polio vaccine.

In 1998, I became a voting member of the Advisory Committee for Immunization Practices (ACIP) to the CDC. The ACIP is the group that recommends how vaccines are used in the United States. My first task was to head the polio-vaccine working group, with an eye toward moving us away from Sabin's live, attenuated polio vaccine and back to Salk's killed-virus vaccine. For several reasons, it was clearly time to abandon Sabin's vaccine in favor of Salk's. First, every year eight to ten children suffered polio from Sabin's vaccine. Second, other countries had completely eliminated polio without every resorting to Sabin's vaccine. Third, advances in protein chemistry and protein purification during the past forty years assured the production of an inactivated vaccine that wouldn't contain any residual live poliovirus.

I assumed that the switch from Sabin's vaccine back to Salk's would be a no-brainer. But I was wrong. Several members of the committee, including D. A. Henderson, who had been short-listed for the Nobel Prize due to his successful efforts to eliminate smallpox from the face of the earth, worried that the inactivated polio vaccine might still contain live virus. Henderson and others continued to point to the Cutter Incident as proof—an example of being made more nervous rather than wiser by experience.

In 2000, the United States switched back to Jonas Salk's inactivated polio vaccine. Now, with the elimination of polio caused by the polio vaccine, Americans can finally say that their country is polio free.

———

Anne Gottsdanker graduated from Reed College in Portland, Oregon, where she majored in biology and philosophy. In 1967, with several of her friends, she chained herself to a draft board to protest the Vietnam War. Later, pursuing an interest

in psycholinguistics, she received a master's degree from the
University of California, Santa Barbara. She married, had
two children, and became a biology teacher. But Gottsdanker
remains permanently paralyzed and uses a wheelchair. "I
probably settled for things in my life that I might not have,"
she said. "When you have a disability that people can see,
they're a lot less accepting—that never changes."

——

One final legacy of the Cutter Incident remains. In the fall of
2020, during the development of vaccines against COVID-19,
several newspapers and online news sources referred to the
Cutter Incident. The manufacturing process for coronavirus
vaccines was called Warp Speed, a name given by Dr. Peter
Marks, a fan of the television series *Star Trek* and the head
of the Center for Biologics Evaluation and Research (CBER),
part of the FDA. To save time, the Trump administration had
charged Warp Speed with making hundreds of millions of
doses of coronavirus vaccines from several different pharmaceutical companies before large, placebo-controlled trials
had proved that the vaccines were safe and effective. The reason the Cutter Incident was back in the news was that Warp
Speed mimicked the polio-vaccine program in 1955—which
could have reasonably been called Warp Speed One—when
five pharmaceutical companies also mass-produced polio
vaccines prior to the completion of the Thomas Francis field
trial that would show that the vaccine was safe and effective.

For people deciding whether or not to receive a coronavirus vaccine, the Cutter Incident should be kept in mind. Even
when large, prospective, placebo-controlled studies of more
than thirty thousand people have shown that coronavirus
vaccines are safe and effective, that assumes no problems will

be encountered with mass production. The Cutter Incident wasn't the Cutter Incident; it was the Scale-Up Incident. So, it's reasonable to pay attention to the first few million doses of any new coronavirus vaccines that are mass-produced using vaccine strategies and materials that have never been used before.

CHAPTER 7

CLARENCE DALLY

X-Rays

ON JULY 2, 1881, JAMES GARFIELD, THE TWENTIETH PRESIDENT OF THE United States, walked into the Sixth Street train station in Washington, DC. Traveling to Massachusetts to speak at his alma mater, Williams College, he was accompanied by his sons James and Harry, Secretary of State James Blaine, and Secretary of War Robert Todd Lincoln, whose father had been assassinated sixteen years earlier. Lying in wait was Charles Guiteau, a failed politician who believed that Garfield had slighted him. Guiteau pulled out a British Bulldog revolver and fired twice. One bullet grazed Garfield's shoulder; the other entered his lower back, lodging behind his pancreas. Knowing that the bullet was a source of infection, doctors probed deeply into Garfield's back with their hands. Unable to find the bullet, they commissioned Alexander Graham Bell to rig up a metal detector that, owing to the distracting metal springs in the bed, was useless. Garfield lived for several weeks before dying from the festering wound.

On September 6, 1901, William McKinley, the nation's twenty-fifth president, was greeting well-wishers at the Pan-American Exposition in Buffalo, New York. As he stuck out his hand, Leon Czolgosz, an anarchist, reached under a handkerchief for a .32 caliber revolver and fired twice. One bullet struck McKinley in the shoulder; the other pierced his stomach. As had been the case with James Garfield, the surgeon failed to locate the bullet. One week later, the president died of a massive bacterial infection. Following McKinley's death, Theodore Roosevelt became the twenty-sixth president of the United States.

On October 14, 1912, Theodore Roosevelt traveled to Milwaukee, Wisconsin, to deliver a speech on behalf of the Progressive (or Bull Moose) Party. When Roosevelt was walking out of the Gilpatrick Hotel on his way to the Milwaukee Auditorium, John Schrank, a saloonkeeper who had been stalking Roosevelt for weeks, pulled out a .38 caliber Colt Special and shot Roosevelt once in the chest, missing his heart. The bullet was deflected by a steel eyeglass case and a fifty-page manuscript jammed into Roosevelt's jacket. Unlike Garfield and McKinley, Roosevelt had the advantage of a diagnostic test that revealed a bullet lodged between his third and fourth ribs. Remarkably, before going to the hospital—and with blood seeping into his shirt—Roosevelt delivered a ninety-minute speech. "Ladies and gentlemen," he began, "I don't know whether you fully understand that I have just been shot. But it takes more than that to kill a Bull Moose!"

Before the procedure that located the bullet in Roosevelt's chest was invented, doctors probed blindly with their fingers to find cysts, tumors, stones, and foreign objects. Diagnosing broken bones was also a grim affair. A textbook written in 1796 instructed doctors to "grasp the limb, move it around, and listen for scraping or grating sounds caused by the broken pieces." With the availability of the new test, all of this blind searching could end. Now, doctors could see fractures or foreign bodies or tumors or pneumonia or ovarian cysts or kidney stones or bullets. They could see inside the patient!

———

Wilhelm Conrad Röntgen was a professor of physics at the University of Würzburg in Germany. On the evening of Friday, November 8, 1895, when he was fifty years old, Röntgen sent an electric current through a glass tube that contained a partial vacuum. This was nothing new. Physicists had been studying the effects of electricity on gases for decades. The difference was that Röntgen was about to discover something that was previously unimaginable.

The tube that Röntgen studied was called a Crookes tube. (Today, Crookes tubes are called cathode-ray tubes and are used in radar displays and televisions.) Other physicists had shown that Crookes tubes could produce Crookes rays, which are high-speed electrons capable of traveling a few inches outside the tube. After leaving the tube, this concentrated beam of electrons could cause certain chemicals to emit light (fluoresce) on contact. Preparing for a later experiment, Röntgen had coated a sheet of paper with a chemical known to fluoresce when struck by high-energy electrons. He set the paper aside. But when Röntgen fired up the Crookes tube with an electric current, he noticed a flicker of light, a faint shimmering, on the sheet that was several feet away. This didn't make sense. The sheet was too far from the tube to have been affected by the Crookes rays. Also, Röntgen had covered the tube with thick, black cardboard. This, too, should have prevented the rays from leaving the tube. Röntgen's first step was to make sure that no other source of light was entering the room. After tightening all the blinds and shades, the faint glow persisted. When he turned off the electric current in the Crookes tube, the glow disappeared. When he turned the electricity on, it reappeared. The conclusion was inescapable; these unusual, far-reaching, high-energy rays were coming from inside the tube.

For the next seven weeks, Röntgen ate and slept in his laboratory, ignored his teaching duties, barred his assistant from the lab, and rarely saw his family. His wife, Anna Bertha, worried that he was losing his mind. Röntgen found that not only did these new rays penetrate thick cardboard, causing the screen to fluoresce, they also penetrated a one-thousand-page book, a double pack of playing cards, a single piece of tinfoil, thick blocks of wood, several sheets of hard rubber, and a plate of aluminum. They didn't, however, penetrate lead. Then he put his hand between the tube and the chemical-coated sheet and saw something that was spectral, inexplicable, otherworldly: the bones of his hand projected onto the screen! Röntgen, who was blind in one eye following a childhood illness, was also colorblind, which is why he never noted the greenish-yellow color of the fluorescence in his logbook. He questioned what he had just seen. "Was it fact or illusion?" he later recalled. "I was torn between doubt and hope." In a letter to his longtime friend, physicist Ludwig Zehnder, he wrote, "I had spoken to no one about my work. To my wife I merely mentioned that I was working on something about which people would say, when they found out about it, 'Roentgen has surely gone crazy.'"

On December 22, 1895, Röntgen brought his wife into the lab. This time, instead of using a sheet of paper coated with a fluorescent chemical, he used a glass photographic plate to capture the image permanently. He asked his wife to put her hand between the tube and the glass plate and to keep it there for fifteen minutes. When she saw the bones of her hand, as well as an outline of the signet ring on her finger, she screamed, "I have seen my own death!" This was the response to the world's first permanent radiograph.

Fifty days after the discovery, Röntgen submitted his findings to the Würzburg Physical Medical Society, which published his handwritten article under the title "Über eine neue Art von Strahlen" ("On a New Kind of Ray"). In his article, Röntgen named these new rays for the mathematical symbol of the unknown: X-rays.

Röntgen was an immediate celebrity. Streets were named after him. People stopped him on the street. Statues were built in his honor. On December 10, 1901, he won the first Nobel Prize in Physics "in recognition for the extraordinary services rendered by the remarkable rays subsequently named after him." (Many countries still refer to X-rays as Roentgen rays.) Although it is now routine, Röntgen didn't deliver a lecture at the Nobel Prize ceremony. He humbly accepted his prize and sat down.

———

On December 30, 1895, only two days after Röntgen had published his paper, a woman came to Queen's Hospital in Birmingham, England, with a needle embedded in her hand. Dr. J. R. Ratcliffe and his colleague, John Hall-Edwards, made an X-ray of her hand, gave it to her, and told her to bring it to her surgeon the next morning. This was the first X-ray obtained to assist in a surgical operation.

On January 4, 1896, one week after Röntgen had published his findings, the Berlin press picked up the story. The next day, news of Röntgen's discovery flew by telegraph across the globe. The *New York Times* wrote, "Probably never before has the entire scientific world been simultaneously aroused to such a pitch of excitement as that caused by the recent discovery."

Within months, the news of Röntgen's discovery had been translated into almost every language in the world. By the end of 1896, fifty essays and a thousand articles about these miraculous new rays had been published. At first, the new technique was called skiagraphy (shadow graphing); later it was called radiography.

The new technology debuted in America on February 3, 1896, when physicists at Dartmouth College took an X-ray of the wrist of Eddie McCarthy, a boy who had fallen while ice-skating on the Connecticut River. The X-ray revealed a broken ulnar bone in his wrist.

Within five years of Röntgen's invention, X-rays were considered essential for clinical care. In 1900, at Pennsylvania Hospital in Philadelphia, 1.3 percent of all patients were X-rayed; by 1925, it was 25 percent. In 2009, a poll conducted by the Science Museum in London named the discovery of X-rays as more important than the discoveries of penicillin, computers, motorized cars, the telegraph, and the DNA double helix.

———

People were at once mystified, frightened, intrigued, and duped by the new technology. Hucksters claimed that X-rays could turn ordinary metal into gold. A newspaper in Cedar Rapids, Iowa, wrote, "George Johnson, who has been experimenting with the X-rays, in three hours [changed] a cheap piece of metal worth about 13 cents to $153 worth of gold. The metal so transformed has been tested and pronounced pure gold." Psychics claimed that X-rays could see bodily spirits before they were liberated by death; physicists brought scientific equipment to séances to see if X-rays could explain how spiritualists read minds or saw into the future. In New York City, it became fashionable for women to make X-ray portraits. "A good many women have their hands taken just for fun, perhaps for a family souvenir," wrote the *New York Times*. X-ray machines were set up in opera houses, on streetside stands, and in department stores. Comic books advertised "X-ray glasses." Superman had X-ray vision.

Mostly, however, the public feared a loss of privacy. They worried that tax authorities would use X-rays to detect hidden treasures or that highway robbers would exploit X-rays to empty pockets. A London newspaper advertised "X-ray-proof underclothing." On February 19, 1896, only two months after Röntgen had published his discovery, Assemblyman Reed of Somerset County, New Jersey, introduced a bill "prohibiting the use of X-ray opera glasses in theaters."

Many in the public worried that scientists had become the sorcerer's apprentice, releasing a magic over which they had no control. Although the public's fears about loss of privacy were silly and ill-founded, they were right to be afraid. One unanticipated consequence, as it turned out, was far worse than they could have imagined.

———

The first victim of X-rays' unseen harms was a young man named Herbert Hawks, a recent graduate of Columbia College. In August 1896, Hawks got a job at Bloomingdale's in New York City demonstrating an X-ray machine. Working two to three hours a day, Hawks took X-rays, sometimes putting his head next to the machine so that customers could see the bones of his jaw. After four days, his hands swelled, his skin was sunburned, his knuckles ached, his fingernails stopped growing, and the hair on his face and temples fell out. By the end of two weeks, the skin on his hands sloughed off, his eyelashes fell out, his eyelids swelled, and his vision blurred. Hawks tried wearing gloves, covering his head with petroleum jelly, and wrapping his hands in tinfoil, all without effect. After two weeks, he stopped working. Doctors and scientists debated the cause of Hawks's symptoms. Could it have been the electric sparks from the high-voltage generator? Or the ultraviolet radiation coming from the tube? Or the fluorescent chemicals used to capture the X-ray images?

Other stories followed:

- A Boston Crookes-tube manufacturer named G. A. Frei reported that the skin on his hands and those of a coworker had turned red, hardened, and sloughed off and that their fingernails felt as if they had been "pounded by a hammer."

- A scientist from the University of Minnesota described how his entire forehead had become an "angry sore" and that his mouth was so blistered he could eat only small amounts of liquid food.

- A thirty-year-old man named William Levy, who had been shot by an escaping bank robber, had lived for ten years with a bullet lodged in his skull. He wanted a surgeon to remove it, but not before being subjected to a fourteen-hour X-ray exposure. Within a day, his entire head had blistered; after a few days, his lips were cracked and bleeding, his right ear had doubled in size, and the hair on the right side of his head had fallen out.

By the end of the year, X-ray burns were front-page news in virtually every prominent electrical, medical, and scientific journal.

No one, however, paid a greater price than the men and women on the front lines of this new technology: radiologists and radiology technicians, most of whom saw themselves as noble warriors, "martyrs to science," in their quest to save lives with X-rays.

In November 1896, Walter Dodd, a founding father of radiology in the United States, suffered severe skin burns on both hands. Within five months, the pain was "beyond description" and his face and hands were visibly scalded. When the pain kept him awake at night, Dodd paced the floor of Massachusetts General Hospital with his hands held above his head. In July 1897, he received the first of fifty skin grafts, all of which failed. Bit by bit, his fingers were amputated. Dodd waited as long as he could before amputating his little finger because, as he said, "I needed something to oppose my thumb."

On August 3, 1905, at the age of forty-six, Elizabeth Fleischmann, the most experienced woman radiographer in the world, died from X-ray-induced cancer after a series of amputations. Fleischmann had gained international renown for her X-rays of soldiers in the Philippines during the Spanish-American War. Upon her death, almost every major newspaper published eulogies about "America's Joan of Arc." Another martyr to the science of radiology.

John Hall-Edwards, the British photographer who had participated in filming one of the first X-ray images and an ardent X-ray advocate, believed that X-rays weren't causing any of these problems—until he could deny it no longer. In January 1897, at the dawn of X-ray technology, Hall-Edwards, who had been exposed to hours of X-rays every day, wrote, "We have heard so much about the effect of the X-rays upon the skin. This I think must be due to some idiosyncrasy of the operators, for although I have myself been experimenting daily for the last eleven months, I have failed to notice anything of the kind." By 1899, now an honorary fellow of the Royal Photographic Society, Hall-Edwards was starting to change his mind. "Continued and protracted exposure to the rays at varying distances from the tubes has an effect upon the hands which, although unpleasant, is not dangerous. It interferes with the growth and nutrition of the nails. The skin around the roots of these [nails] becomes red, irritable, and cracked; the nails themselves [become] thin and brittle. Most constant workers suffer in this way." In 1904, finally convinced of X-rays' permanent, severe harms, John Hall-Edwards strongly urged young workers to take every possible precaution before it was too late; the pain in his hands had now progressed to the point that it was "as if bones were being gnawed away by rats." By 1906, his left arm was useless and carried in a sling. In 1908, when cancerous growths were found, the arm was amputated just below the elbow and the fingers of his right hand were removed.

By 1911, more than fifty cases of X-ray-induced cancers had been reported, primarily among radiologists. In 1926, the *New York Times* reported the seventy-second operation on Johns Hopkins Hospital radiologist Frederick H. Baetjer, a physician whose research had already cost him eight fingers and one eye. "Despite the suffering he had undergone in the interest of science," the *Times* reported, "Baetjer plans to continue his work as long as he lives, fingers or no fingers." In honor of the selflessness of these early practitioners, Dr. Percy Brown wrote a book titled *American Martyrs to*

Science Through the Roentgen Rays. Brown died from X-ray-induced cancer twenty years later.

Historian Bettyann Kevles describes a 1920 professional gathering of radiologists, where so many attendees were missing hands and fingers that when the chicken dinner was served no one could cut their meat.

===

Elizabeth Fleischmann wasn't the first American to die from experimenting with X-rays. She was the second. The first was a man named Clarence Dally, who would become famous because of the notoriety of his boss: Thomas Alva Edison.

In 1879, more than a decade before Wilhelm Conrad Röntgen discovered X-rays, Edison invented the electric light bulb, which produced visible light by passing an electric current through a carbon or tungsten filament. After reading about Röntgen's discovery, Edison had a better idea. Instead of producing light with electricity through a filament in a glass bulb, he would bounce X-rays off of fluorescent chemicals painted on the inside of a Crookes tube. Edison hoped that this new light bulb would be brighter and less expensive than the one he had invented. The technician in charge of the project was Clarence Dally.

In 1896, when he was twenty-four years old, Dally joined the Edison lab in West Orange, New Jersey. His job was to make Crookes tubes and test them for efficiency. He did this by placing his left hand between the tube and the detecting screen. (Dally was right-handed.) If the image of his hand on the screen was sharp, then he knew that the tube was well-made. If not, he made the proper adjustments. During his eight years in the Edison lab, Dally made hundreds of Crookes tubes and took thousands of pictures of his hand, often spending as many as ninety hours a week working with the harmful rays.

At first, Dally lost his hair, eyebrows, and eyelashes. His face wrinkled, resembling a man in his sixties, not thirties. A deep, painful ulcer appeared on the back of his left hand, which became severely inflamed. Dally then stopped testing the apparatus with his left hand, using his right hand instead. The pain in his hands eventually became so constant, so unremitting, and so severe that he could sleep only if he put them in pails of water. By 1901, he had had seven surgical operations—involving more than 150 pieces of skin taken from his leg and grafted onto his left hand—all without success. In 1901, a cancer developed on his left arm, which was amputated. In 1903, four fingers were removed from his right hand. Then his right arm was amputated. When Dally could no longer work, Edison kept him on the payroll. On October 2, 1904, at the age of thirty-nine, Clarence Dally died of cancer, leaving behind a wife and two young children.

When Dally died, Thomas Edison abandoned X-rays. Regarding his idea for an X-ray-powered light bulb, he said, "I could make the lamp all right. But when I did so I found that it would kill everybody who would use it continuously." Years later, when Edison was asked about Dally's experiments, he remarked, "Don't talk to me about X-rays. I am afraid of them." When confronted with a persistent toothache, Edison insisted that the dentist pull the tooth outright instead of subjecting him to a dental X-ray.

Although Elizabeth Fleischmann and Clarence Dally were among the first to die from working with X-rays, they were far from the last. In 1936, in the garden of St. George's Hospital in Hamburg, Germany, a memorial was erected to the early pioneers of X-rays who had lost their lives to their work. Inscribed on the monument are 169 names from fifteen countries with the citation: "To the Röentgenologists and Radiologists of All Nations, Doctors, Physicists, Chemists, Technical Workers, Laboratory Workers, and Hospital Sisters who gave their lives in the struggle against diseases. . . . Immortal is the glory of the work of the dead." Antoine Béclère, a French radiologist, spoke at the dedication: "These noble martyrs did not speak the same language, did not belong to the same country, they were of different races and religions [but] all devoted to the mission of fighting, at the peril of their lives, the same enemies . . . without fear that this weapon was double-edged and would one day wound and kill them." By 1959, the number of X-ray deaths had increased from 169 to 359.

———

Before answering the question about whether these early X-ray tragedies were preventable, it might help to understand exactly what X-rays are.

We'll start at the beginning.

All matter is composed of atoms. The paper on which this book is written, the ink used to print the words, the eyes used to read the words, and the brain used to process the words are all composed of atoms. Atoms are like tiny solar systems. At the center of the atom (the nucleus) are two different types of particles: protons and neutrons. Protons are positively charged; neutrons are neutral. Therefore, the nucleus of atoms is positively charged. Surrounding the nucleus is a cloud of electrons, each of which is negatively charged. Atoms don't like to be positively or negatively charged. They like to be neutral. So, if there is one proton in the nucleus, then there will be one electron orbiting around it. An atom with one proton and one electron is called hydrogen. Helium has two protons and two electrons. Oxygen has eight protons and eight electrons. And so on through the 118 atoms (or elements) that make up the periodic table of elements.

Atoms combine to form molecules. For example, two hydrogen atoms and one oxygen atom combine to form the molecule water (H_2O). Eight carbon atoms, ten hydrogen atoms, four nitrogen atoms, and two oxygen atoms combine to form the molecule caffeine ($C_8H_{10}N_4O_2$).

If an atom loses an electron, it isn't neutral anymore. Because electrons are negatively charged, losing an electron means the atom is now positively charged. If an atom gains an electron, it becomes negatively charged. Atoms that are positively or negatively charged are called ions. This is where X-rays come into the story. X-rays are ionizing, meaning they turn neutral atoms into ions.

X-rays and visible light have one thing in common. Both travel as waves of energy called electromagnetic waves. The difference between these two different kinds of electromagnetic waves is the frequency or size (wavelengths), as well as the energy of the waves. X-rays have shorter wavelengths and higher energies than visible light—much higher energies. For this reason, X-rays can travel through skin and, if unobstructed, come out the other side to expose an X-ray film. If the X-rays encounter air (such as in the lungs), they travel through the body and expose the X-ray film, turning it black. If X-rays encounter bone, they don't travel through the body. Rather, they are absorbed by the bone. In this case, the X-ray film appears white, or unexposed. Bones are white and lungs are black on a chest X-ray. Visible light, on the other hand, isn't powerful enough to travel through the body. Light from the sun casts a shadow because it doesn't penetrate the skin.

Visible light and X-rays are two of the many different kinds of waves on the electromagnetic spectrum. The least dangerous are radio waves, microwaves, infrared light, and visible light, none of which are ionizing. The most dangerous are X-rays, gamma rays, and high-frequency ultraviolet rays, all of which are ionizing. (It's the ionizing ultraviolet rays from the sun that cause skin cancer.)

The reason that X-rays are dangerous and visible light isn't is that ionizing radiation can disrupt molecular bonds. The most important molecule that is disrupted by X-rays is DNA, the blueprint of life. Once DNA is damaged, cells can become cancerous. Cancers caused by ionizing radiation have been most dramatically demonstrated by the after-effects of nuclear power plant disasters (such as Chernobyl) and atomic bombs (such as those dropped on Hiroshima and Nagasaki at the end of World War II).

———

Combine the power of X-rays with the primitive nature of early X-ray machines and it's easy to understand why so few radiologists who attended that professional meeting in 1920 could cut their chicken. And why so many subsequently died from cancer.

The first X-ray machines were bulky, loud, sparking, smelly devices easily prone to mishap and injury. To some extent, every X-ray machine that was sold in the early 1900s was unreliable. Static generators used as a source of electricity were finicky and susceptible to changes in humidity; Crookes tubes were fragile, poorly focused, and often produced rays of varying wavelengths; and glass photographic plates were breakable and required developing skills that few physicians possessed.

Typically, to create an adequate image, patients were brought close to the tubes, which occasionally exploded, showering them with molten glass. Because the electric generators were large and immobile, the patient and the radiologist often had to contort themselves at odd angles to take the picture. Patients were slung headfirst over chairs or strapped to boards or suspended in chains over the machine, where they had to remain motionless for hours during the exposure. Techniques to immobilize patients ranged from large, heavy sandbags to anesthesia.

For the operator of the X-ray machine, getting an adequate image was a nightmare. The amount of vacuum in Crookes tubes varied widely. No two were alike. Radiologists had to test each tube for quality, which meant taking X-rays of their own hands, usually several times, before they could be assured of an adequate image.

Recently, scientists in the Netherlands unearthed an X-ray machine similar to those used in the early 1900s. They found that the amount of radiation emitted was 1,500 times greater than that to which patients are exposed today. Also, where exposure times in the early 1900s ranged from ten minutes to several hours, exposure times today are about twenty milliseconds (thousandths of a second). And, unlike the early days of X-ray technology, medical technicians now leave the room before turning on the machine.

Between 1896 and 1930, tens of thousands of radiologists, technicians, and patients suffered burns, hair loss, and bone pain; thousands lost fingers, hands, and arms; and hundreds died from cancer.

Was the harm caused by these early X-rays preventable?

The strongest evidence that the decades-long use of unshielded X-ray machines, unshielded patients, and unshielded radiologists and technicians was entirely preventable was provided by the one person who never developed skin burns, hair loss, cancer, or any of the other gruesome side effects of prolonged exposure to X-rays: Wilhelm

Conrad Röntgen. In his first paper, published in January 1896, Röntgen clearly described how these new rays didn't penetrate lead. Indeed, while studying X-rays, Röntgen always encased his Crookes tube in a lead-lined box, protecting himself from the harmful rays.

In 1898, two years after Röntgen's discovery, lead-lined rubber gloves became commercially available. By 1905, full protective gear that included aprons, gloves, and hoods was routinely used in Britain. In 1908, England's John Hall-Edwards published a list of rules, including a small aperture for the X-ray beam, a recommended distance from the Crookes tube by the operator, and routine use of an opaque apron and lead spectacles. That same year, in a London hospital, operators stopped using their own hands to check the quality of the image and all tubes were encased in lead boxes.

In the 1910s and 1920s, every element of the X-ray apparatus underwent at least one major improvement. Thin-walled glass Crookes tubes yielded to the more rugged and adjustable hot-cathode tubes; glass photographic plates were replaced by film; near universal electrification rendered static generators obsolete; alternating current allowed for higher voltages; diaphragms blocked stray radiation and sharpened images; and shortened exposure times became routine.

By the end of the 1920s, X-ray machines had finally been transformed from an unreliable, smoking spectacle into an unobtrusive piece of clinical equipment. In 1928, the International Congress of Radiology finally stepped forward to regulate this industry, agreeing on a system to measure radiation and proposing safety guidelines for workers.

Today, five billion X-ray exams (which include standard X-rays, CT scans, mammograms, and dental X-rays) are performed every year. The thirty thousand radiologists, two hundred thousand dentists, and two hundred thousand X-ray technicians who perform them are at no greater risk of cancer than the general public.

———

In the early 1900s, as Americans learned more and more about the dangers of this new technology, X-ray machines disappeared from opera houses, street-side stands, and department stores. They did, however, persist in one popular venue. During the 1930s, 1940s, and 1950s, shoe stores used X-ray machines to assist in fittings. At the peak of this fad, about ten thousand shoe stores promoted their X-ray machines. It wasn't until the 1970s that they were banned from shoe stores in thirty-three states and strict regulations in the remaining seventeen made their use impractical. The last remnant of this dangerous practice was unearthed in 1981 in the shoe section of a department store in Madison, West Virginia.

PART III

SERENDIPITY

IN 1901, WHEN HENRY TAYLOR MADE DIPHTHERIA ANTISERUM, HE knew that the horse from which he had harvested the serum had died from tetanus. His error caused the deaths of thirteen children.

In 1937, when Harold Watkins chose diethylene glycol as a solvent for the world's first antibiotic, he should have known that he had made a product that could cause kidney failure. This potentially fatal side effect had been published in a scientific journal six years earlier. Because of Watkins's ignorance, 105 people died; thirty-four were children.

These kinds of preventable tragedies aren't limited to the distant past. For example, in 2014, when health-care workers in Syria inadvertently diluted a measles-mumps-rubella (MMR) vaccine with a muscle relaxant instead of a saltwater solution, fifteen children stopped breathing and died. This wasn't the fault of the vaccine or the vaccine maker.

Medical research and medical care are human endeavors. And humans are flawed. These tragedies did nothing to advance knowledge or make for better products. Diphtheria antisera, sulfanilamide tablets, and MMR vaccines continued to be made in the same manner as before the disasters.

Some tragedies, on the other hand, have led directly to medical products that save lives. In the next two chapters—about cancer treatments and gene therapies—you'll see how disasters that caused the deaths of eleven children in Boston and one teenager in Philadelphia gave birth to medical products that are today saving thousands of lives.

CHAPTER 8

ELEVEN UNNAMED CHILDREN

Chemotherapy

SECOND ONLY TO HEART DISEASE, CANCER IS THE SINGLE BIGGEST killer in the United States, accounting for one of every four deaths. In 2021, doctors will diagnose 1.8 million new cancers and six hundred thousand people will die from the disease.

In children, the most common type of cancer is leukemia—specifically, acute lymphoblastic leukemia (ALL). In the 1940s, ALL was a death sentence. Today, because of chemotherapy, most children are cured. The road to successful chemotherapies, however, has been littered with tragedy, none greater than one that caused the immediate deaths of eleven children in 1946.

———

The first cancer chemotherapy was based on a mass poisoning that occurred during World War I. On July 12, 1917, German commanders released a toxic gas on unsuspecting British troops near the small town of Ypres, Belgium. At first, soldiers noticed a thick, shimmering cloud around their feet and a strange peppery smell in the air. Unlike other forms

171

of chemical warfare, however, this particular gas diffused through leather, rubber, and cloth; gas masks were useless. For days, the gas hung over the battlefield. Soldiers initially suffered painful, ulcerating blisters. Then, after the gas penetrated into their lungs, they coughed up blood. Some were permanently blinded. Because the Germans couldn't control the wind, soldiers on both sides were affected. The poison gas was made by reacting a dye, thiodiglycol, with hydrochloric acid. Because it smelled like mustard, it was called mustard gas. When the war finally ended, mustard gas had wounded 1.2 million soldiers and killed more than a hundred thousand.

Those who survived mustard-gas attacks later developed severe anemia, requiring monthly blood transfusions. They were also prone to recurrent, lingering, and sometimes fatal infections. In 1919, one year after World War I ended, two American pathologists, Helen and Edward Krumbhaar, performed autopsies on seventy-five soldiers who had been killed by mustard gas. They found that the gas depleted the bone marrow, where red blood cells, white blood cells, and platelets are made. They also found that lymph nodes, another source of white blood cells, had shrunk. The Krumbhaars published their findings in 1919. No one noticed. Specifically, no one recognized that if mustard gas could eliminate white blood cells and shrink lymph nodes, maybe it could also eliminate cancers of the bone marrow (leukemias) and cancers of the lymph nodes (lymphomas).

Fast forward to World War II.

In the early evening of December 2, 1943, a fleet of Luftwaffe planes struck the harbor near Bari, a fishing town on the southern coast of Italy. Fifteen American ships sank and eight were severely damaged. One ship, the SS *John Harvey*, which was carrying millions of gallons of gasoline, vaporized in a massive explosion. Also—unknown to many of its crew—the ship was carrying seventy tons of mustard gas. Although the Geneva Protocol had outlawed the use of poison gas in 1925, both Winston Churchill and Franklin Delano Roosevelt worried that the Germans might not comply, so they developed their own arsenal.

Of the more than six hundred men floating in the Adriatic Sea off the coast of Bari that day, eighty-three died in the first week. During the next few months, one hundred more died. The US Army sent Lieutenant Colonel Stewart Alexander, a physician with an expertise in chemical warfare, to investigate. Alexander soon realized that the deaths could only have been caused by mustard gas. Similar to the findings of Helen and Edward Krumbhaar, Alexander found that survivors suffered bone marrow depletion and shrinkage of their lymph nodes. When Alexander was ready to issue his report, however, Churchill ordered that the mustard-gas burns be reported as "dermatitis [skin inflammation] due to enemy action." Churchill didn't want the world to know that the Allies had violated the Geneva Protocol.

———

At the time of the explosion off the coast of Bari, the US Chemical Warfare Service was an undercover operation within the Office of Scientific Research and Development (OSRD). After Lieutenant Colonel Alexander submitted his report, the OSRD offered a contract to two Yale University pharmacologists, Louis Goodman and Alfred Gilman, to investigate mustard gas as a possible treatment for cancer. First, Goodman and Gilman converted the gas to a liquid by substituting a nitrogen atom for a sulfur atom. The new agent was called nitrogen mustard. Next, they tested its effects on rabbits and mice with lymphomas, finding that this

novel drug shrunk tumors and prolonged lives. Finally, they
convinced a friend and surgeon, Dr. Gustaf Lindskog, to give
the drug to a forty-eight-year-old New York man with lymphoma. The work was done in secret. It wasn't until January
19, 2011, almost seventy years later, that the medical records
of Lindskog's patient were unearthed from on off-site storage
facility at Yale. The patient was code-named "JD."

JD was born in Poland in 1894. When he was eighteen years old, he immigrated to the United States, where he
worked in a ball-bearing factory. In August 1940, a severe
form of lymphoma invaded the entire right side of his neck.
He could barely open his mouth, turn his head, swallow, or
sleep. In February 1941, he was referred to the Yale Medical
Center for radiation therapy. After two weeks of daily radiation, he improved. But the improvement was short-lived.
By August 1942, he had trouble breathing, couldn't eat, and
had lost a substantial amount of weight. On August 27 at 10
a.m., JD became the first person in history to receive a medicine to treat cancer. Every day, for ten consecutive days, he
received an injection of nitrogen mustard. After the fifth dose,
his tumor regressed; finally, he was able to move his head
and eat. One month later, however, his tumor came back,
necessitating another three-day course of nitrogen mustard;
again, the response was short-lived. So, he received a six-day
course, without effect. On December 1, 1942, ninety-six days
after he had received his first dose of nitrogen mustard, JD
died. Because this was a covert operation run by the OSRD,
the phrase "nitrogen mustard" never appeared in his medical
chart. Instead, doctors referred to it as "substance X."

The first paper describing nitrogen mustard's effects on
cancer wasn't published until 1946, four years after JD was
treated. On October 6, 1946, the *New York Times*, under the
headline "War Gases Tried in Cancer Therapy," wrote, "The

possibility that deadly blister gases prepared for wartime use may aid victims of cancer will be investigated by the Army Chemical Corps' Medical Division." Nitrogen mustard had provided the first ray of hope in the fight against cancer.

The modern age of chemotherapy had begun.

——

Encouraged by the results with lymphoma, scientists and doctors turned their attention to the most heartbreaking of all cancers: childhood leukemia. In the 1940s, the average life span of a child diagnosed with ALL was less than four months. It was the rare child who lived for a year; no child lived past fourteen months. Most of those diagnosed with ALL were between three and seven years of age.

Children with ALL have bone marrows that are packed with cancerous white blood cells, replacing red blood cells and platelets. Hundreds of thousands of these cancerous cells then pour out of the bone marrow into the bloodstream. (The word leukemia comes from the Greek meaning "white blood.") Because these cancerous white blood cells are dysfunctional, children are prone to severe infections. In the mid-1940s, no drugs were available to treat ALL. Children were given blood transfusions for their anemia and antibiotics for their infections. Otherwise, they were placed in distant wards to die in peace. Their deaths, however, were anything but peaceful.

As cancerous white blood cells packed bone marrows, bones became brittle and easily broken. These fractures, called pathological fractures, were excruciatingly painful. Children with ALL lost weight, were constantly fatigued, and suffered enlarged livers that made it difficult to eat and enlarged spleens that easily ruptured. In an age before platelet transfusions, children often experienced massive blood loss following trivial cuts or falls. Mouth ulcers, bleeding gums, and swollen lymph nodes were just a few of the many frightening symptoms. Through all of this, parents were forced to stand back and watch helplessly as their children died slowly in front of them.

The first person to push back against this cruelest of diseases was Dr. Sidney Farber. Farber was born in Buffalo, New York, the third of fourteen children. In 1923, he graduated from the State University of New York with a plan to attend medical school. In the mid-1920s, however, Jews were often refused admission. So, Farber, who spoke fluent German, went to Europe instead, attending medical school at the University of Heidelberg, where he excelled. Later, he transferred to Harvard Medical School, graduating in 1927. After studying pathology at the Peter Bent Brigham Hospital (now Brigham and Women's Hospital), he became the first pathologist based at a children's hospital. Between 1947 and 1948, Farber was made pathologist in chief at Children's Hospital in Boston and professor of pathology at Harvard Medical School. During his career, Farber published more than 270 scientific papers; his book *The Postmortem Examination* remains a classic. Today, Farber's name has been immortalized in one of the world's most famous cancer centers: The Dana-Farber Cancer Institute in Boston, part of Harvard Medical School. In short, Sidney Farber was the last person anyone would have imagined would have been at the center of one of the worst chemotherapy tragedies in history.

—

The origins of the tragedy can be traced to the work of Dr. Lucy Wills. In 1928, Wills, having recently graduated from the London School of Medicine for Women, traveled to Bombay to investigate an unusual form of severe anemia in women. She assumed that the anemia was the result of a nutritional deficiency. At the time, two forms of nutritional anemias had been described. The first was iron-deficiency anemia. But Wells found that these women had plenty of iron in their diet. So that wasn't it. The second was called pernicious anemia, which could be cured by eating raw liver, an excellent source of vitamin B_{12}. Wills found that eating livers also cured the anemia in these Indian women. But it wasn't the B_{12} that did it. It was something else. At first, that something else was called the Wills factor. Soon it was called folic acid, which is also contained in dark-green, leafy vegetables such as spinach, kale, asparagus, and broccoli. (The word folic is derived from the Latin *folium*, meaning "leaf.") As it turned out, folic acid, also known as vitamin B_9, is critical for the synthesis of DNA (the blueprint for cell growth and reproduction) and RNA (responsible for making the proteins necessary for the cell's survival). In short, folic acid is a cellular growth factor, which should have made it the one thing a doctor would never want to give to a child who needed to stop rapidly growing cancerous cells.

In 1938, Daniel Laszlo, a Viennese physician, teamed up with Richard Lewisohn and Rudolf and Cecilie Leuchtenberger at Mount Sinai Hospital in New York City to study the role of nutritional factors in treating cancers. Six years later, in 1944, the team published its first paper, titled "'Folic Acid' a Tumor Growth Inhibitor." To the surprise of everyone, the group found that folic acid inhibited the growth of tumors in seven mice. Then, they found that a "folic acid concentrate" inhibited tumors in 117 mice. Ten more experiments showed that it inhibited tumors in 364 mice. In the paper, which was published in the prestigious *Proceedings of the Society for Experimental Biology and Medicine*, the Mount Sinai team always put the words "folic acid" in quotes. They assumed that the antitumor activity found in their liver and yeast extracts was caused by the vitamin. But because folic acid hadn't yet been synthesized or isolated in pure form, they hedged their bets as to what was really treating the mice.

On January 12, 1945, the Mount Sinai team published a second paper, titled "The Influence of 'Folic Acid' on Spontaneous Breast Cancers in Mice." Again, the paper was published in a prestigious journal (*Science*) and again the researchers put folic acid in quotes. This time, they found that what they believed was folic acid had caused the complete regression of spontaneous breast cancers in thirty-eight mice.

Sidney Farber read these papers with interest—at last, something that might relieve the suffering of children with ALL. In 1946, Farber called an old friend, Yellapragada Subbarow, who worked for Lederle Laboratories, a pharmaceutical company in upstate New York. Farber asked Subbarow to provide a purified form of folic acid. Unlike the researchers at Mount Sinai, Farber wasn't going to use liver or yeast extracts as a source of folic acid. He was going to give the children something purer, safer, and more powerful. By the summer of 1946, Farber had what he needed. During the next few weeks, he administered pure folic acid to eleven children with ALL. In one patient, the number of cancerous white blood cells circulating in the bloodstream doubled. In another, leukemic infiltrates burst out of the bone marrow and onto the skin. In every case, folic acid hastened their deaths. It's not easy to treat a disease that has a mortality rate of 100 percent and a life span of only a few months and make it worse. But Sidney Farber had done exactly that. Farber later remarked that the bone marrows of children treated with folic acid were packed with more leukemic cells than he had ever seen.

Farber never published a report about what had happened to these children. Instead, he referred to the event in subsequent publications as an "acceleration phenomenon," a euphemism for precipitating the rapid deaths of eleven children. Pediatricians at Boston Children's Hospital and across the country were furious at Farber's hubris.

On November 8, 1946, the same year that Sidney Farber gave folic acid to eleven children with ALL, the Mount Sinai researchers published a paper in *Science* explaining the confusion. In this third paper, the researchers made it clear why they had been putting folic acid in quotes. It wasn't folic acid after all. It was the opposite of folic acid, something later called folic acid antagonists. Again, the Mount Sinai team studied mice with spontaneous breast cancers, finding that, whereas untreated mice developed new tumors and lung metastases, mice treated with folic acid antagonists had a complete regression of their tumors. Folic acid antagonists work by blocking the effects of folic acid, depriving DNA and RNA of a needed nutrient. Unlike folic acid, folic acid antagonists aren't growth promoters; they're growth inhibitors.

Although mice aren't people, this was a seminal paper—the first to show that folic acid antagonists treated cancers and that folic acid worsened them. The question at this point was when did Sidney Farber read this third study from the Mount Sinai group? Was it before or after he had hastened the deaths of eleven children with ALL? Farber injected children with folic acid in September 1946, two months before the Mount Sinai study showed its harmful effects. So, assuming that Farber hadn't seen the Mount Sinai data before they were published—either because they hadn't been presented at a national meeting or because they hadn't been discussed among colleagues—he can't be held accountable for the tragedy of his making.

The question at this point was would these folic acid antagonists have an anticancer effect in people. On June 3, 1948, Farber and his colleagues at Boston Children's Hospital published a paper in the *New England Journal of Medicine* that has become one of the most cited, most referenced research papers in the history of cancer therapy. It was titled "Temporary Remissions in Acute Leukemia in Children Produced by Folic Acid Antagonist, 4-Aminopteroyl-Glutamic Acid (Aminopterin)." The cumbersome title understated the results, which were remarkable.

Beginning on September 6, 1947, Farber had injected sixteen moribund children with ALL with a folic acid antagonist called aminopterin. Ten of the sixteen went into remission; leukemic cells no longer packed their bone marrows and soon disappeared from their bloodstreams. Livers, spleens, and lymph nodes shrunk. White blood cell counts returned to normal. Bleeding stopped and appetites returned. Children went back to school and played with their friends. A two-year-old boy named Robert Sandler got up and walked for the first time in two months. Another boy started to run and play after lying in bed for seven months. Two of the patients were still alive sixteen and twenty-three months later. Although everyone in the trial eventually relapsed and died, for the first time a childhood cancer had been treated successfully with a medicine. The modern-day form of aminopterin, called methotrexate, remains an important chemotherapy for ALL.

———

Of interest, the first person to receive this new chemotherapy wasn't among the sixteen children described in Sidney Farber's *New England Journal of Medicine* paper. Rather, it was an older man described by Dr. Richard Lewisohn from the Mount Sinai group on September 5, 1947, at the Fourth International Cancer Congress in Saint Louis. During his presentation, Lewisohn never mentioned the man's name, for a reason.

Standing at the podium, Lewisohn described a fifty-two-year-old man who, during the fall of 1946, noticed that his voice was growing increasingly hoarse. Then he experienced severe pain behind his left eye. One doctor, who thought the man had a tooth abscess, had recommended pulling three teeth, which didn't help. The man worsened, unable now to speak or swallow. An X-ray revealed a large tumor at the base of his skull. Radiation therapy did little to lessen the growth of the tumor or spare him the excruciating pain. At this point, he was referred to Lewisohn, who treated him with a folic acid antagonist for six weeks, with clear improvement. The tumor shrank and the man regained most of the eighty pounds he had lost the previous year. On June 13, 1948, the man felt well enough to put on his old New York Yankees uniform—the one with the number 3 on the back—and attend the twenty-fifth anniversary of the opening of Yankee Stadium. His name was Babe Ruth.

———

Since Sidney Farber's study of sixteen children in 1948, progress in treating ALL has been remarkable.

In 1965, inspired by multidrug therapy to treat tuberculosis, physicians employed the first combination therapy for cancer (prednisone, vincristine, methotrexate, and 6-mercaptopurine). The use of combination drugs prevented mutation and relapse of the cancer cells. The result was the first long-term remissions for ALL. Physicians cautiously talked about a cure. By 1968, ALL cure rates hit 50 percent. By the late 1970s, 70 percent.

Today, using drugs like dexamethasone, vincristine, L-asparaginase, daunorubicin, and methotrexate, 95 percent of children with ALL go into remission and 90 percent are cured—an amazing achievement. But like all these early achievements, it came with a price, perhaps best summed up by the first person to receive a folic acid antagonist. "I realized that if anything was learned about that type of treatment, whether good or bad, it would be of use in the future to the medical profession and maybe for a lot of people with my same trouble." Babe Ruth was a hero to children, on and off the baseball diamond.

CHAPTER 9

JESSE GELSINGER

Gene Therapy

JESSE GELSINGER, THE SON OF PATTIE AND PAUL GELSINGER, WAS born in Tucson, Arizona, on June 18, 1981. His birth and early life were unremarkable. Jesse was a picky eater, refusing meat and dairy products, choosing potatoes and cereal instead. No one made anything of it. Neither Pattie nor Paul realized that his eating pattern was a clue to a rare disease.

In January 1984, when he was two and a half years old, Jesse became tired, sleepy, and listless, unable to rouse himself to participate in any activity. Pattie took him to the doctor, who said that Jesse needed more protein. Pattie Gelsinger started forcing her son to drink milk and eat bacon and peanut butter sandwiches. As it turned out, this was exactly the worst thing she could have done.

Two months later, Jesse woke up, parked himself in front of the television to watch his favorite cartoons, and fell asleep. Unable to wake him, Pattie took Jesse to the hospital, where doctors found an abnormally high level of ammonia in his bloodstream. Jesse had a rare genetic disease called ornithine transcarbamylase (OTC) deficiency, which affects

183

about one in eighty thousand children in the United States every year.

This unusual enzyme deficiency explained why Jesse had been avoiding protein. When the body breaks down protein for energy, one product is ammonia—no different than the ammonia used to scrub floors. Ammonia is highly toxic. If levels in the bloodstream get too high, ammonia travels to the brain, causing coma, brain damage, and eventually death. To rid the body of ammonia, the liver makes five enzymes that convert it to urea, which is then eliminated from the body in the urine. One of those enzymes is OTC. Newborns with OTC deficiency typically slip into a coma in the first few days of life. Half die before they are a month old; the other half are dead before their fifth birthday. All of these children lack the single gene that makes the OTC enzyme.

When his diagnosis became clear, Jesse was put on a low-protein diet and given medicines to help rid his body of ammonia. In a sense, Jesse was lucky. Some of the cells in his liver made a little bit of OTC, which explains why he didn't die in the first few years of life. If he stuck to his low-protein diet and took his medicines—which amounted to thirty-two pills a day—he could live a long life. But a low-protein diet and dozens of pills are a tough regimen for a young boy. In 1991, when he was ten years old, after a weekend-long binge on high-protein foods, Jesse slipped into a coma and was again rushed to the hospital.

Jesse's disease was frustrating because—apart from an inability to make the one enzyme needed to rid his body of ammonia—his liver was perfectly normal. So, although a liver transplant would have cured him, it would have been much simpler to find a way to give Jesse the one gene that made the one enzyme that he lacked. In other words, gene therapy.

In September 1998, Jesse and Paul Gelsinger learned of just such a gene-therapy program at the University of Pennsylvania in Philadelphia. Researchers had put the gene that Jesse lacked into a harmless virus, which, acting as a Trojan Horse, could deliver the OTC gene into Jesse's liver cells, avoiding the need for a liver transplant.

On December 22, 1998, three months after the Gelsingers had learned about Penn's gene-therapy program, Paul Gelsinger returned home to find his son curled up on the couch, vomiting uncontrollably. By the time Paul got him to the hospital, Jesse was in a coma; his ammonia level six times above normal. By Christmas, Jesse was in the intensive-care unit and on a breathing machine. For two days, he lingered in and out of consciousness. Although it was touch and go, Jesse recovered. Chastened, he never missed another dose of his medicines again. But now, more than ever, Jesse Gelsinger wanted to participate in Penn's gene-therapy program. The man in charge of the program was Dr. Jim Wilson.

——

In the 1970s, Jim Wilson obtained both an MD and a PhD from the University of Michigan. While at Michigan, he studied an unusual syndrome called Lesch-Nyhan, which, like OTC deficiency, is rare, affecting only about one in 380,000 people in the United States. Also similar to OTC deficiency, the symptoms of Lesch-Nyhan syndrome appear in the first few years of life when, to the abject horror of parents, children engage in self-injurious behavior, banging their heads against the wall and chewing on their lips, tongue, and fingers until they bleed. Patients also exhibit a unique symptom called coprolalia: an uninterrupted stream of uncontrolled, involuntary cursing—like a scene out of the movie *The*

Exorcist. No other disease has this symptom. Wilson was the first to prove that Lesch-Nyhan syndrome was caused by a deficiency of a single gene. How easy it would be, then, to fix it. Just give children the gene that they needed, and all of these frightening symptoms would disappear. Wilson committed himself to finding a way to provide children with the genes they were missing.

In 1980, Wilson opened the journal *Science*. There, he found an article that changed his life. Written by a pair of Stanford biochemists, Richard Mulligan and Paul Berg, it was titled "Expression of a Bacterial Gene in Mammalian Cells." Mulligan and Berg had accomplished something that up to that point had seemed unimaginable. They had taken a gene from a bacterium called E. coli (which normally lives in the intestines), put it into monkey cells, and showed that they could actually coax these cells into making a bacterial protein. Mulligan and Berg had found a way to rewrite the DNA of monkeys! This was the beginning of gene therapy. When Richard Mulligan left Stanford to work at MIT, Jim Wilson left Michigan to join him. Years later, energized that he could do in human cells what Mulligan had done in monkey cells, Wilson was chosen to head the Institute for Human Gene Therapy at the University of Pennsylvania.

The first genetic disease that Wilson tackled at Penn was something called familial hypercholesterolemia. Circulating in the bloodstream of every human are two different types of cholesterol: low-density lipoprotein (LDL) cholesterol, otherwise known as "bad cholesterol," and high-density lipoprotein (HDL) cholesterol, or "good cholesterol." The reason that bad cholesterol is bad is that it can damage the lining of arteries, especially ones that supply the heart and brain. People with familial hypercholesterolemia lack the one gene that makes the one protein (called LDL-binding protein) that attaches to bad cholesterol and eliminates it from the body. As a consequence, they have a high incidence of strokes and heart attacks. Like OTC deficiency and Lesch-Nyhan syndrome, familial hypercholesterolemia is a rare disease, affecting about one out of every million people.

Wilson's first subject was a twenty-nine-year-old, French-speaking seamstress and part-time bank teller from Quebec who suffered from the disease. She had had her first heart attack when she was sixteen years old. When she was twenty-six, she had undergone bypass surgery, where one of her damaged heart arteries was replaced with a vein from her leg. Two of her brothers, who also suffered from familial hypercholesterolemia, had died of heart attacks in their twenties. To provide this woman—who refused to give her name to the press or allow herself to be photographed—the gene that she needed, Wilson had to perform a major operation. On June 5, 1992, a surgeon removed 15 percent of the woman's liver, separated and grew her liver cells in sterile, plastic dishes in the laboratory, and infected the cells with a harmless virus that contained the gene that she lacked. The surgeon then reinfused one billion of these virus-infected liver cells back into the woman's bloodstream. Wilson hoped that at least some of these genetically modified cells would settle back into the woman's liver and make LDL-binding protein. It worked. About 3 to 5 percent of her liver cells began to make the needed protein, pulling bad cholesterol out of her bloodstream. As a consequence, bad cholesterol levels decreased and good cholesterol levels rose. Speaking through an interpreter, she said, "I feel I can do more physical activity, like skiing, dancing, and other social activities. I'm certainly going to live until ninety years of age."

In 1994, Jim Wilson published a paper in a medical journal describing this groundbreaking experiment. The next day, the front page of the *New York Times* hailed the event as "the first effort to reverse an inherited disease permanently by altering the genetic makeup of a patient's cells." "This shows that the principle of gene therapy is sound and that it can work," said Wilson. Dr. John Kane, director of the Lipid Clinic at the University of California, San Francisco, was stunned by the result. "This is a landmark experiment," he enthused. "It's the Kitty Hawk of gene therapy." In the *New York Times* article, Wilson mentioned that he was also interested in patients with high levels of ammonia in their bloodstream caused by a specific enzyme deficiency. The enzyme Wilson was talking about was OTC, the same one that Jesse Gelsinger was missing.

What the *New York Times* reporter failed to realize, however, was that Jim Wilson wasn't the first researcher to successfully alter a patient's cells to produce a missing gene. Before Wilson's gene-therapy breakthrough, French Anderson had been called the "father of gene therapy" and honored at a ceremony in George H. W. Bush's White House. Anderson, however, would later become someone that his fellow scientists, the media, and the public would try to forget.

A brilliant student and track star, Anderson graduated from Tulsa Central High School in 1954, Harvard College in 1958, Cambridge University in 1960, and Harvard Medical School in 1963. During his career, Anderson published more than four hundred research papers, forty editorials, and five books. He received dozens of awards, including five honorary doctorate degrees. In 1994, four years after his famous but now forgotten experiment, he was hailed by *Time* magazine as a "hero of medicine." In 1998, he was inducted into the Oklahoma Hall of Fame along with singer Reba McEntire.

On September 14, 1990, two years before Jim Wilson injected genetically modified liver cells into the Quebec woman with familial hypercholesterolemia, French Anderson treated a four-year-old girl named Ashanthi DeSilva, who suffered from severe combined immunodeficiency disease (SCID). Known to the public as "bubble-boy disease," SCID means that a child is born without a functioning immune system. As a consequence, sufferers are often overwhelmed by infections before they reach their first birthday. Children with SCID—like those with Lesch-Nyhan syndrome, familial hypercholesterolemia, and OTC deficiency—lack only a single gene. To give Ashanthi the gene that she needed, Anderson removed some of the immune cells from her blood, infected them with a virus containing the missing gene, and, in a hospital room at the NIH in Bethesda, Maryland, reinfused them into her bloodstream. During the next two years, Ashanthi received ten more such infusions. Eventually, her immune system returned to normal; no longer did she have to suffer life-threatening infections. Twelve years later, her immune system remained intact. Now in her thirties, Ashanthi is married and working as a writer and journalist. Anderson described her progress in medical journals in 1995, 1996, and 2003. Three years later, he disappeared from public view.

On July 30, 2004, one year after he had published Ashanthi DeSilva's final progress report, French Anderson was arrested. Two years later, he was convicted on three counts of lewd acts on a child and one count of continuous sexual abuse. The victim, who was the daughter of a colleague in Anderson's lab, was ten years old when the abuse began; Anderson was sixty. During the trial, a tape was played of Anderson admitting to the girl—who, unknown to Anderson, had been wearing a wire—that his actions were "indefensible" and "just evil." He was sentenced to fourteen years in prison. Twelve years later, at the age of eighty-one, he was released on parole. Anderson now wears an ankle bracelet that allows authorities to monitor his whereabouts.

━━

Although the basis of gene therapy is straightforward—replace a missing or defective gene with one that works—the dragon in the cave is the vector that carries the gene into cells. Typically, the vectors have been viruses. When French Anderson used gene therapy to treat immune deficiency or Jim Wilson used it to treat high cholesterol, they chose the same type of virus to supply the missing gene: a retrovirus. The best-known retrovirus is human immunodeficiency virus—the cause of AIDS. Most retroviruses, however, are benign; they don't cause disease. The advantage of using retroviruses is that they directly insert the missing gene into DNA. This means that when the cell divides, each of the daughter cells will contain the new gene. The other major advantage of these harmless retroviruses is that they fly below the radar of the immune system, which doesn't recognize them as foreign and try to destroy them.

When Jim Wilson arrived at the University of Pennsylvania, he abandoned retroviruses in favor of a different virus vector: adenovirus. Wilson was concerned about one potential problem with using retroviruses, which European researchers later learned the hard way. (We'll get to that in the epilogue.) Adenoviruses, however, had their own problems. Unlike retroviruses, adenoviruses *can* cause disease—specifically, infecting the nose, throat, eyes, lungs, bladder, and intestines. Also, unlike retroviruses, adenoviruses can be recognized by the immune system. This means that the virus can be eliminated before it enters cells and provides the needed gene. Adenoviruses can also destroy the cells that they enter, which is why they cause so many symptoms. To use adenoviruses, researchers had to find a way to disable the virus so that it could express the missing gene without hurting the cell or alerting the immune system. As it turned out, this was much harder than anyone had imagined.

Beginning in 1994, Wilson and his colleagues published scientific papers describing how they had disabled adenovirus so that it could express the needed gene in the liver, fly below the radar of the immune system, and leave liver cells unharmed. The gene that they put into their adenovirus vector was the one that Jesse Gelsinger lacked: OTC. Wilson's team found that their disabled adenovirus vector containing the OTC gene protected mice against the infusion of large quantities of ammonia. Then they did the same experiment in baboons. Again, the disabled adenovirus vector did what it was supposed to do. Now, they were ready to try it on people. They submitted an investigational new drug license to the FDA and were granted permission to get started. (Several disabled adenoviruses are also being used to make COVID-19 vaccines.)

The year was 1997. At this point, Wilson was a gene-therapy rock star. With an annual budget of $25 million and a staff of 250 researchers, he now headed the largest academic gene-therapy program in the world.

Wilson first tested his disabled adenovirus vector containing the OTC gene on adults who, like Jesse Gelsinger, had a partial OTC deficiency. The bioethicist at the University of Pennsylvania who advised the trial, Dr. Arthur Caplan, thought this would be better than testing it on severely affected babies. Following the tragedy that was to come, this was one of several decisions that came under fire.

The Penn team first divided volunteers into six groups of three participants each. The first group received one billion virus particles per pound of body weight, injected directly into the artery that supplies the liver. Dr. Steven Raper was the surgeon who injected the OTC-gene-containing adeno-viruses. Patients were carefully monitored for evidence of liver damage. If the first group of participants didn't have serious problems, investigators would give the next group a higher dose. The sixth and final group, the one in which Jesse Gelsinger would soon find himself, received the highest dose: three hundred billion viral particles per pound of body weight. Some of the participants in the first five groups ex-perienced flu-like symptoms; mild, short-lived abnormalities of the liver; and a transient decrease in platelets. None of these patients, however, experienced serious side effects. Also, about half of the participants showed a modest increase in OTC levels—a promising sign.

In April 1999, Paul Gelsinger informed Jesse's doctor in Tucson that they were interested in participating in Wilson's experimental program at Penn. That same month, Jesse re-ceived a letter accepting him into the program. On June 18, 1999, Jesse Gelsinger, his father, stepmother, and three sib-lings—PJ, age nineteen; Mary, age fifteen; and Anne, age four-teen—boarded a plane to Philadelphia. It was Jesse's birthday. That night, they had a party at the home of Paul's brother. On June 22, Jesse and Paul met with the trial coordinators to review the consent form. The details contained in this form would later become the focus of a series of federal fines and civil lawsuits. During the next few days, Jesse visited Pat's famous cheesesteak restaurant, the Betsy Ross House, Inde-pendence Mall, the Liberty Bell, and the Rocky statue in front of the Museum of Art, where Jesse's picture was taken—a picture that would later be circulated as a cautionary tale in newspapers across the country.

—

On Monday morning, September 13, 1999, Jesse Gelsinger was taken to the interventional-radiology suite at the Hospital of the University of Pennsylvania and strapped to a table, where a doctor threaded a catheter into the artery that supplied his liver. At 10:30 a.m., Dr. Steven Raper slowly injected one ounce of the disabled adenovirus into Jesse's artery, finishing two hours later. That night, Jesse spiked a fever to 104.5 degrees. He also had back pain, a headache, and some dizziness when he sat up. Raper wasn't surprised. The same reaction had occurred in several other patients. Paul Gelsinger called from Tucson and spoke briefly with his son, exchanging "I love yous" at the end. It was the last time they would speak.

On Tuesday morning, September 14, the whites of Jesse's eyes had become yellow, or jaundiced. This meant that Jesse's liver wasn't functioning properly. Unlike the fever, which had abated, jaundice had not been seen in any of the other patients who had received the disabled adenovirus vector. By midafternoon, twenty-four hours after the injection, Jesse lapsed into a coma. With ammonia levels in his bloodstream rising, doctors put Jesse on a dialysis machine (which mimics the function of the kidneys). Although his ammonia level quickly returned to normal, he remained comatose. The high level of ammonia in his bloodstream apparently wasn't the cause of his coma.

On Wednesday morning, September 15, Paul Gelsinger took a red-eye flight from Tucson back to Philadelphia, arriving at 8 a.m. Jesse was critically ill, on a breathing machine, and in a deep coma. Raper was beside himself, deciding finally to put Jesse on an ECMO machine. ECMO, which stands for extracorporeal membrane oxygenation, is used during major heart surgery to act as an artificial lung. Raper hoped that the ECMO machine would give Jesse's progressively stiffening lungs a chance to rest. In 1999, ECMO had been used in fewer than a thousand patients, only half of whom had survived. At first, ECMO seemed to be working. Then Jesse's kidneys shut down.

On Thursday, September 16, Hurricane Floyd battered the East Coast. Mickie Gelsinger, Jesse's stepmother, flew into Philadelphia just before the airport closed. Pattie, Jesse's mother, was being treated in a psychiatric facility in Tucson, unable to travel. That night, Paul couldn't sleep. So, he walked the half mile to the hospital to see his son. Jesse's face was bloated beyond recognition. His eyes and ears were completely closed off by the swelling. "How can anyone survive this?" said Paul.

On Friday, September 17, Jesse Gelsinger was brain dead. Paul called a chaplain to perform a bedside service before doctors withdrew life support. The chaplain anointed Jesse's head with oil and recited the Lord's Prayer. Doctors fought back tears. With life support withdrawn, Steven Raper put a stethoscope to Jesse's chest and officially pronounced him dead. "Goodbye, Jesse," he said. "We'll figure this out."

Paul Gelsinger was forgiving. "I went so far as to tell the doctors that I didn't blame them, that I would never file a lawsuit." All that would soon change. "Little did I know what they really knew," Paul said later.

In 1989, ten years before Jesse Gelsinger died, the FDA hadn't approved a single trial of gene therapy. By September 1999, at the time of Jesse's death, it had approved ninety-one. Gene therapy, it seemed, was on the verge of major breakthroughs. By the end of 2000, however, gene therapy had become a tale of men playing God and suffering the consequences.

On a clear Sunday afternoon in early November 1999, atop the jagged peak of Mount Wrightson, 9,500 feet above Tucson, Jesse Gelsinger was laid to rest. Among the two dozen mourners were Jesse's father, mother, stepmother, brother, three doctors, and a handful of friends. Paul talked about Jesse's love of motorcycles and professional wrestling. All agreed that he had a ready wit and kind heart. Steven Raper, the doctor who had injected Jesse with the fatal dose of adenovirus, pulled a small blue book from his pocket and read from an elegy by Thomas Gray. "Here rests his head upon the lap of Earth," recited Raper, "a youth to Fortune and Fame unknown. Fair Science frowned not on his humble birth." Jesse's ashes were then scattered into the canyon below. "Every realm of medicine has its defining moment, often with a human face attached," wrote Sheryl Gay Stolberg for the *New York Times* in an article titled "The Biotech Death of Jesse Gelsinger." "Polio had Jonas Salk. In vitro fertilization had Louise Brown, the world's first test-tube baby. Transplant surgery had Barney Clark, the Seattle dentist with the artificial heart. AIDS had Magic Johnson. Now gene therapy has Jesse Gelsinger."

Researchers, politicians, and clinicians would spend the next ten years trying to figure out what had happened to Jesse Gelsinger. They were all trying to answer one question: Was his death preventable?

Jesse's death was immediately reported to the FDA and the Recombinant DNA Advisory Committee (RAC) at NIH. It took almost two weeks, however, for the mainstream press to find out about it. On September 29, 1999, Rick Weiss and Deborah Nelson of the *Washington Post*, under the headline "Teen Dies Undergoing Experimental Gene Therapy," wrote, "An 18-year-old Arizona man with a rare metabolic disease has died while participating in a controversial gene-therapy experiment, marking the first death attributed by doctors to a burgeoning field of research." The reason that the *Post* used the word "controversial" would soon become clear.

On the same day that the article appeared in the *Washington Post*—twelve days after Jesse Gelsinger's death—federal officials sent letters to more than one hundred researchers working with similar adenovirus vectors, asking them to report any evidence of trouble. At the time, thousands of patients had been treated with various types of gene therapies.

In December 1999—three months after Jesse Gelsinger's death—the RAC held a meeting at NIH to find out what had gone so horribly wrong. For two days, with his peers on a stage in front of him and a large audience of the press and public behind him, Jim Wilson answered questions. And for two days, he said that he didn't know why Jesse had died when eighteen people who had received the same disabled adenovirus vector before him hadn't. Ron Crystal of Cornell said that he had administered a virus vector similar to Wilson's more than one hundred times and seen only one serious reaction, which was short-lived. No one, however, was satisfied. Abbey Meyers, president of the National Organization for Rare Disorders and former RAC member, said, "Everybody has to share in the guilt of what's happened here."

In January 2000—one year after Jesse Gelsinger's death—the FDA suspended the University of Pennsylvania's gene-therapy program and began an investigation into sixty-nine other programs.

Nine months later, in September 2000, the Gelsinger family filed a lawsuit against the Penn researchers, settling out of court for an undisclosed amount. Also that month, Donna Shalala, US secretary of health and human services, said, "This appalling state of affairs is unacceptable." In an

article published in the *New England Journal of Medicine*
titled "Protecting Research Subjects—What Must Be Done,"
Shalala vowed to put systems in place to prevent future tragedies and to pursue legislation where researchers would now
be held personally responsible for their actions, including
penalties up to $250,000 per investigator and $1 million per
institution.

In April 2002—three years after Jesse Gelsinger's
death—Jim Wilson stepped down as director of the Institute
for Human Gene Therapy at the University of Pennsylvania.
The institute later disbanded and Wilson never again headed
a clinical trial.

In February 2005—six years after Jesse Gelsinger's
death—the US Department of Justice brought civil charges
against Jim Wilson, Steven Raper, and the University of
Pennsylvania. The Department of Justice argued that Wilson,
Raper, and Penn had violated the federal False Claims Act by
failing to obtain proper informed consent and making false
statements in grant applications and progress reports. The
University of Pennsylvania agreed to pay a fine of $517,496.
Paul Gelsinger wasn't satisfied. "This judgment lets everyone
off the hook," he said.

After the lawsuits, fines, institutional changes, and re-
vised federal guidelines, many believed that justice had been
served—that gene therapy would finally be held to a standard
that would prevent similar tragedies in the future. But, as it
turned out, all of these measures had nothing to do with why
Jesse Gelsinger had died.

First, federal officials were angry that Jesse Gelsinger's
ammonia level was seventy at the time of his admission to the
program, when the protocol approved by the FDA stated that
subjects with ammonia levels greater than fifty would not
be admitted. But Jesse's death, as was later learned, wasn't

caused by his high level of ammonia, which often fluctuated into those ranges, at the start of the trial.

Second, federal officials were angry that the consent form that Jesse signed never mentioned the two monkeys that had died following injection of Wilson's disabled adenovirus vector. Both of these monkeys, however, had received an adenovirus vector that wasn't nearly as disabled as Jesse's vector. And both monkeys had received a dose of vector that was almost twenty times greater than that given to Jesse. Also, the monkeys had died from inflammation of the liver (hepatitis), which wasn't what had killed Jesse, as Wilson would later learn. The consent form did warn that, given what had happened to the monkeys, liver dysfunction could be a problem, stating, "It is even possible that this inflammation could lead to liver toxicity or failure and be life-threatening."

Third, lawyers accused Wilson of a financial conflict of interest. In 2010, Professor Robin Fretwell Wilson, in a law journal article titled "The Death of Jesse Gelsinger: New Evidence of the Influence of Money and Prestige in Human Research," wrote that Jesse's death was "arguably the most famous conflict-of-interest case in medicine." The article noted that while he was at the University of Michigan, Jim Wilson had founded a biotech company called Genovo, which had the right to market any or all of his discoveries. By 1999, when Steven Raper injected Jesse Gelsinger with the adenovirus vector, Genovo had provided more than $4 million a year to Wilson's institute—a substantial portion of its budget. Wilson and his family had a 30 percent equity stake in Genovo, and the University of Pennsylvania had a 3.2 percent stake. But, despite the public outcry, Genovo hadn't sponsored Jesse Gelsinger's experiment. Also, to avoid the appearance of a conflict of interest, Wilson wasn't allowed to make any clinical decisions regarding participants in the trial. Those decisions were made by Raper, who had no financial stake in the outcome. Unfortunately, Paul Gelsinger refused to believe that anything other than greed and the undue influence of Big Pharma had killed his son, later writing, "What is wrong is that a growing, ambitious minority of researchers and institutions have compromised their ethics for profits and prestige, mostly as a result of the industry's inappropriate financial influence over them and our government."

Fourth, the bioethicist involved in the program, Arthur Caplan, was criticized for insisting that the trial include only adults who had little chance of benefiting. But Caplan, as well as other ethicists unaffiliated with the Penn program, argued that adults were better able to provide informed consent. Also, it would have been difficult to recognize life-threatening events from gene therapy in babies who were sick and dying from the disease.

Ten years after Jesse's death, Jim Wilson wrote, "My deepest regret is that a courageous young man who agreed to participate in this clinical trial with the hope of making life better for others with this disease lost his life in the process. The [outcome] was unanticipated and not predicted based on the preclinical and clinical data available at the time." In the intervening years, Wilson and his team figured out exactly what had happened to Jesse Gelsinger.

———

Guangping Gao is a microbiologist who was an associate director of Penn's gene-therapy program at the time of Jesse Gelsinger's death. "We were all shocked and lost," he recalled. But Jim Wilson wouldn't let his team quit. "Wilson told us we all had work to do," said Gao. "As professionals, we had to get beyond the emotions of that moment. We had to focus on doing everything we could—every sample, every hypothesis—to figure this out." The now-abandoned board-room of the Translational Research Lab at Penn remains filled with artifacts from Wilson's ill-fated gene-therapy trial. Books such as *Building Public Trust* and *Biosafety in the Laboratory* sit on the shelf. But the most interesting artifacts are the names of two products of the immune system scrib-bled on the whiteboard: IL-6 and TNF-alpha. The first tells you everything you need to know about what had happened to Jesse Gelsinger.

After Jesse Gelsinger died, Wilson's team reviewed every scrap of clinical and laboratory data from the participants who had received the disabled adenovirus vector. One thing stood out. After the injection, almost all had had a bump in their blood levels of interleukin-6 (IL-6)—a weapon used by the immune system to fight infections. The difference between Jesse and the other participants was that his IL-6 level—apart from being much higher than the others—never decreased, remaining extraordinarily high until the moment he died. The clinical term for this reaction is a "cytokine storm." (IL-6 is a cytokine. Cytokines are proteins made by the immune system that can cause a disease that looks like sepsis, an over-whelming bacterial infection. IL-6 is also one of the cytokines that are elevated in many of those who die from the novel coronavirus, SARS-CoV-2.)

Why did Jesse continue to produce massive levels of IL-6 while other participants didn't? Between 2001 and 2005, Wilson and his team performed a series of studies in mice and monkeys to answer this question. They found that Jesse's disabled adenovirus traveled to the outer rims of the spleen, entered specific immune cells called macrophages, and induced massive amounts of IL-6. Worse, the toxicity curve was steep. The vector appeared to be perfectly safe un-til researchers reached an amount of virus that was too great;

then, suddenly and unexpectedly, it was deadly. This is exactly what Wilson had seen in the clinical trial where—after more than two years, proceeding slowly and cautiously, increasing the dose a little bit at a time—one patient suddenly fell off a cliff. The question remained, however, why the participant before Gelsinger—a girl in her late teens who had received the same dose of the same adenovirus vector—didn't have the same problem. She, too, initially had high quantities of IL-6 in her bloodstream, but her levels quickly fell off.

In 2003, Steven Raper, in an article for a medical journal describing Jesse Gelsinger's case, wrote, "The experience points to the limitations of animal studies in predicting human responses." He also discussed the "steep toxicity curve" and "subject-to-subject variation" in the trial. "With what I know now," said Wilson in 2009, ten years after Jesse's death, "I wouldn't have proceeded with this study." But Wilson hadn't known. And nothing about the animal model studies or previous experience in people had allowed him to know. It's a little like saying that you never would have bet on a horse to win a race after you found out that it had lost. Of course you wouldn't.

Jesse Gelsinger's death, although unpredictable and therefore unpreventable, would not be in vain.

———

In May 2010, a five-year-old girl named Emily Whitehead from central Pennsylvania was diagnosed with ALL. Although the chemotherapy regimen used to treat ALL is daunting—seven or eight different potentially toxic medicines, some of which are injected directly into the spinal fluid—the prognosis is excellent. About 90 percent of children survive. Soon after the first round of chemotherapy, however, Emily relapsed. After the second round, she developed a severe infection of both legs with "flesh-eating" bacteria that almost required amputation. Sixteen months later, she had a second relapse. Children who relapse early or relapse twice have poor outcomes. Emily had done both. One doctor recommended hospice care.

After the second relapse, Emily's parents heard about a different kind of therapy being developed by a researcher named Carl June at the University of Pennsylvania. Unfortunately, the therapy hadn't been approved by the FDA. Emily's cancer cells were doubling daily. So, Emily underwent a third round of chemotherapy, which enabled her to live three more weeks, still without remission.

Fortunately, by March 2012, when Emily Whitehead arrived at the Children's Hospital of Philadelphia, the FDA had approved Carl June's trial. By this time, all of Emily's organs were packed with cancer cells. June and his team removed some of Emily's functioning immune cells from her bloodstream and reengineered them to kill Emily's cancer cells. Dr. Stephan Grupp, a pediatric oncologist, then infused the cells (called CAR-T cells) into Emily's vein while she sucked on a popsicle. Feeling fine that night, Emily returned with her parents to her aunt's house, where she was given a piggyback ride by her father. The second evening, however, at the exact same moment that Jesse Gelsinger had begun his descent, Emily developed a high fever and had to go back to the hospital, where she was taken to the intensive-care unit. There, in a manner identical to Jesse Gelsinger, Emily's kidneys shut down, she was placed on a breathing machine, and she drifted in and out of consciousness. "We thought she was going to die," said June. "I wrote an email to the provost at the university telling him the first child with the treatment was about to die. I feared the trial was finished. I stored the email in my out-box, but never pressed send."

But Emily didn't die. Aware of what had happened to Jesse Gelsinger a decade earlier, Stephan Grupp ordered a blood test for IL-6. Similar to Jesse, Emily's IL-6 level was one thousand times greater than normal. Unlike in Jesse's case, however, this wasn't 1999. It was 2012, and a product that countered the effects of IL-6, called tocilizumab, was commercially available. Unfortunately, the drug was licensed only for the treatment of arthritis. Grupp asked for and received permission for the off-label use of the drug. It saved Emily's life. Eight days later, on her seventh birthday, Emily Whitehead woke up. "I have never seen a patient that sick get better so quickly," recalled Grupp. Three weeks later, a biopsy of Emily's bone marrow showed that it was completely free of cancer cells. Several years later, Siddhartha Mukherjee—a well-known cancer specialist, author, and speaker—asked Emily whether she remembered coming into the hospital. "No," she said. "I only remember leaving."

In August 2017, the FDA approved CAR-T therapy for people with difficult-to-treat cancers. Thousands of patients have now received CAR-T therapy, and most have also had to receive the drug that treats high levels of IL-6. All owe their lives to the boy who first showed that massive secretion of IL-6 was a problem and to the researcher, Jim Wilson, who figured it out.

———

CAR-T therapy is a breakthrough product. Carl June and Stephan Grupp are heroes. Jim Wilson, on the other hand, has fallen by the side of the road. Far from being the star of gene therapy he was supposed to have been, Wilson became a pariah. No longer was he invited to give lectures or participate in committees. He's refused to give interviews for more than a decade. The popular view is that June and his team succeeded where Wilson had failed because Wilson had broken the rules. Because he was, as one professor of law had stated, unduly affected by "the corrosive influence of financial interests in human subjects research." But a closer look shows that the only difference between the outcomes of Emily Whitehead and Jesse Gelsinger were luck and timing.

The national media has consistently and rightly praised Grupp for his quick thinking in giving Emily the drug that saved her life. Steven Raper, on the other hand, was successfully sued by the Gelsinger family and censured by the FDA and the US Department of Justice. The public and press have assumed that the different outcomes in these two cases were due to differences in the competence of these two physicians. Where Grupp had quickly recognized that Emily's rapid downhill course was caused by the overproduction of IL-6, Raper had only realized it in retrospect. But even if Raper had known what was happening to Jesse at the time, the drug that countered the effects of IL-6 wasn't commercially available in 1999. So, nothing could have been done to treat or prevent it. Indeed, one of the reasons that Grupp recognized what was happening to Emily was that he knew the cause of Jesse Gelsinger's demise—thanks, in large part, to the hard work of Wilson, Raper, and their team following Jesse's death.

Wilson and his team were also vilified for what the media believed was a conflict of interest with Genovo. Had his disabled adenovirus vector become a medical product, Wilson and the University of Pennsylvania would have benefited financially. The media has failed to note, however, that the name of the institution at Penn that houses Carl June's CAR-T program is the Novartis-Penn Center for Advanced Cellular Therapeutics. CAR-T therapy is now a medical product, and the pharmaceutical company Novartis is the beneficiary. Called Kymriah, the product has also financially benefited both Carl June and the University of Pennsylvania. After the success of CAR-T with Emily Whitehead, no one was upset that June and Penn were associated with Novartis. No one talked about the corrosive influence of Big Pharma on human research. And Tom Whitehead, Emily's father, didn't talk about the "ambitious minority of researchers and institutions [that] have compromised their ethics for profits and prestige," as Paul Gelsinger had done. Industry support for these two programs was indistinguishable. One wonders what Tom Whitehead might have said had Emily not survived her therapy.

The FDA had also criticized Wilson's team for what it believed was a failure to be more forthright in the consent form about the side effects of the adenovirus vector—such as fever, mild liver dysfunction, and lowered platelet counts—all of which were transient and of little to no consequence. On the other hand, the CAR-T program, because it has been so successful, remains unscathed. But a closer look shows that these two programs were more similar than different. Emily Whitehead was the first child to receive CAR-T therapy. But she wasn't the first patient; she was the seventh. The first was a sixty-five-year-old retired corrections officer named Bill Ludwig, who suffered from chronic lymphocytic leukemia. On August 3, 2010, doctors injected Ludwig with his first dose of genetically modified immune cells. Two more infusions followed, after which he became critically ill. His lungs, kidneys, and heart all began to fail, and his temperature rose to 105 degrees. ("The nurses threw the thermometers away, thinking that they had been broken," Carl June recalled.) After several weeks in the intensive-care unit, he recovered. Bill Ludwig, like Jesse Gelsinger before him and Emily Whitehead after him, had suffered from the massive overproduction of IL-6. Fortunately, he recovered without having received the drug to treat it (even though the FDA had licensed the drug eight months earlier), went into remission, and nine years later remains cancer free. Were Emily's parents fully informed about the details of Ludwig's near death? And if Emily had died, would her parents, lawyers, and federal regulators have made the same case that was made of the two monkeys that had died following injections with Wilson's disabled adenovirus vectors?

The only difference between the gene-therapy trials of Emily Whitehead and Jesse Gelsinger was that Emily lived and Jesse died. And when someone dies from an experimental therapy, everyone looks for a fall guy. Jim Wilson was that fall guy.

———

Wilson didn't quit. Although his Institute for Human Gene Therapy has disbanded and he no longer performs clinical trials, he proceeded to revolutionize the field of gene therapy by creating a series of viral vectors that Inder Verma, who headed the NIH committee that had investigated Wilson after Jesse's death, called "gutless adenoviruses."

An autopsy study of Jesse Gelsinger found that the disabled adenovirus vector that had caused his death wasn't in his liver only; it was also detected in every other organ in his body, including the spleen, where it had evoked a massive, fatal immune response. Wilson set about the task of making adenoviruses that were far more specific for various organs and far more disabled once they got there. Called adeno-associated viruses (AAV), they appear to have solved the original problems. Wilson has now made more than three hundred of these viral vectors and generously distributed them to hundreds of investigators across the globe. For example, AAV2 targets the retina in the back of the eye and is now an FDA-approved treatment for an inherited form of blindness. AAV9 targets the brain and is now an FDA-approved treatment for an inherited form of paralysis. "The successes happening now are a legacy of Jesse's death," says Wilson. "We had to succeed." By 2018, more than seven hundred active investigational new drug applications had been submitted to the FDA. This renaissance in gene therapy has been due in no small part to the tireless efforts of Dr. Jim Wilson.

====

In 2019, Siddhartha Mukherjee, writing for the *New Yorker* magazine, visited the University of Pennsylvania gene-therapy program. "The facility may as well have been a small monument to Emily [Whitehead]," he noted. "Photographs of her plastered the walls: Emily at eight in pigtails; Emily at nine with a missing front tooth, smiling next to President Obama; Emily at ten holding a plaque." The iconic picture of Jesse Gelsinger standing in front of the Rocky statue, however, was nowhere to be found. A celebration of the successes, but never the failures that made those successes possible.

LIVING WITH UNCERTAINTY

THE COST OF MEDICAL BREAKTHROUGHS DRAMATIZES SEVERAL ADDI-tional themes that are relevant to today's decision-making.

#1: Nature reveals its secrets slowly, grudgingly, and often with a human price. Scientists, clinicians, academicians, and pharmaceutical company executives must stay humble and respect the requisite learning curve that comes with new discoveries.

The development of COVID-19 vaccines was often ac-companied by a disturbing show of hubris. After completion of phase 1 trials, which examined small numbers of volun-teers given different doses of vaccines, some company re-searchers and executives crowed. Moderna (fifteen patients), Pfizer (thirty-five patients), and AstraZeneca (ten patients) claimed that they could now make tens of millions of doses. These bold pronouncements ignored the likely surprises that lay ahead when a handful of recipients gives way to tens of millions of recipients. This lack of humility was especially concerning given that SARS-CoV-2 had already shown it-self to be an elusive, difficult to characterize virus that had provided a number of surprising clinical and pathological problems, not least of which was inflammation of the blood vessels that could damage any organ, including the brain and heart. No other virus had done what this virus was doing. Also, none of the strategies used by these three companies to make a SARS-CoV-2 vaccine had ever been used to make a vaccine before. Surely, a learning curve lay ahead.

My concern about this unprecedented level of hubris comes from personal experience. In 1980, I began work on a virus called rotavirus, which causes fever, vomiting, and diarrhea in infants and young children. Every year in the United States the virus would infect millions of babies, causing about seventy-five thousand to be hospitalized and sixty to die from severe dehydration. In the world, rotavirus killed about two thousand children every day. So, there was a lot of interest in making a vaccine to prevent it.

Unlike SARS-CoV-2, rotaviruses were well known, having first being described as a cause of disease in animals in the 1940s and in people in the 1970s. Veterinarians had been studying rotaviruses in pigs, sheep, horses, lambs, cows, and other species for decades. And clinicians had been studying human rotaviruses for about ten years. By the time I came into the field, researchers had published more than four hundred papers in medical and scientific journals.

In August 1998, Wyeth, in collaboration with scientists at NIH, released the first rotavirus vaccine. Called RotaShield, the vaccine was given by mouth to about one million American babies before it was unceremoniously pulled off the market when it was found to cause a rare, serious side effect called intussusception. Intussusception occurs when one segment of the small intestine telescopes into the next segment and gets stuck. As a consequence, the surface of the intestine, in the face of a critical loss of its blood supply, is severely damaged. This damage can result in massive blood loss—which can be fatal—or in the entrance of bacteria from the surface of the intestine into the bloodstream—which can also be fatal. During the ten months that this vaccine was on the market, many babies were hospitalized with intussusception and at least one baby died.

The most surprising part of this story is that no one had anticipated this problem. Rotavirus and a vaccine to prevent it had been studied in people and animals for more than five decades without a shred of evidence that this problem was possible. Contrast that with the current pandemic, where the virus that causes it had only been studied for about a year before companies began touting their vaccines, and it's easy to see why humility, not hubris, should rule the day.

#2: Although federal guidelines lessen the chance of disasters, they will never eliminate them. Unanticipated tragedies are unpreventable, no matter how many regulations, training programs, fines, and penalties are put in place.

In 2000, one year after Jesse Gelsinger died, researchers in France inoculated ten children with a retrovirus vector (instead of the adenovirus vector that Jesse had received) to treat a genetic disease. The retrovirus vector inserted the missing gene directly into the cell's DNA. Three years later, two of these ten children developed leukemia. As it turned out, the retrovirus had inadvertently activated an oncogene: inherited genes that increase the risk of cancer.

More than a hundred different oncogenes have been identified, each associated with a specific type of cancer. For example, the HER2 oncogene is associated with breast cancer and the c-Myc oncogene with lung cancer. Many people have oncogenes in their DNA, but these genes aren't harmful because they aren't activated. However, if a retrovirus happens to insert itself in front of an oncogene and turns it on, cancer can be the result. In the two children with leukemia, the retrovirus had awakened an oncogene called LMO2, which is associated with leukemia; in October 2004, one of the two children died.

On January 24, 2005, the French researchers reported a third child who had developed leukemia caused by their retrovirus gene therapy, and later a fourth child. When the dust settled, four of the ten children treated with this retrovirus vector had developed leukemia. In response, the FDA suspended all trials in the United States using this particular viral vector.

What's disturbing about this story is that after Jesse Gelsinger died, systems were put in place to make sure that gene-therapy tragedies never happened again. Federal regulators required extensive testing in experimental animals and rigid monitoring plans before putting a novel gene therapy into people. They also issued guidelines on how to write and obtain informed consent. They created a new federal division, the Office for Human Research Protections, to enforce daunting fines and draconian penalties for those who didn't comply. Nonetheless, despite all these efforts to prevent a second gene-therapy tragedy, one had occurred.

#3: Tragedies shouldn't cause people to lose faith in the scientific endeavor. Science lurches forward in fits and starts, but it inevitably moves forward.

The retrovirus-caused-leukemia tragedy offers another lesson—one that is far more hopeful. In response to the leukemia disaster, researchers modified retrovirus vectors to include an "insulator" gene that eliminated the possibility that the virus could activate an oncogene. The protective gene worked. Researchers at St. Jude's Children's Research Hospital in Memphis were able to permanently correct severe immune deficiencies in ten children using this newer, safer retrovirus vector. Years later, none of these children had developed leukemia. Parents can now safely rely on these modified retroviruses to cure single-gene diseases. Yet another breakthrough built on tragedy.

During the Cutter Incident, when more than one hundred thousand children were inoculated with a polio vaccine that contained live poliovirus, tens of thousands were briefly paralyzed, hundreds were permanently paralyzed, and ten were killed. In response, federal regulators shut down the polio-vaccine program for several months until they could figure out what had gone so horribly wrong. When researchers finally did figure it out, better safety tests were put in place and the problem of polio caused by Jonas Salk's polio vaccine disappeared.

A few months after the Cutter tragedy, polio vaccines were put back on the market. Now, parents had a choice to make. They could either trust that federal regulatory agencies had solved the problem, or they could wait a year or so to make sure that the problem didn't recur. The choice to wait, however, wasn't risk free. Poliovirus was still circulating in the community. In 1955, the year of the Cutter disaster, poliovirus paralyzed twenty-nine thousand people, mostly children. In 1956, because of the polio vaccine, that number dropped to 15,000; in 1957, to 5,500; and by 1962, to 900. Those who were paralyzed between 1955 and 1962 hadn't been vaccinated in part because some parents were still more frightened by the vaccine than the disease, resulting in a bad choice.

In response to the Elixir Sulfanilamide disaster, the FDA immediately recalled Massengill's product. Sulfanilamide preparations made by other companies, however, remained on the market. Early in the investigation, federal regulators and scientists argued about the cause. Was the sulfanilamide itself causing the deaths or was it something else? During this time, some people might have chosen to forgo sulfanilamide on the chance that their infections would heal on their own, rather than risk the possibility of kidney failure and death caused by sulfanilamide. Again, no risk-free choice was available. When it became clear that the problem with Elixir Sulfanilamide had nothing to do with the drug and everything to do with the solvent in which the drug was suspended, people again had a choice. Those who chose to wait for more information made the riskier choice.

In response to the diphtheria antitoxin tragedy in Saint Louis, the Tetanus Board of Inquiry, in collaboration with the police, found that antiserum from a horse with tetanus had been mixed with antiserum from an uninfected horse. At the time, tests for detecting tetanus in antiserum were extraordinarily sensitive. The problem wasn't in the test; it was in the dishonest actions of a worker who had mislabeled the vial. Sadly, within a few weeks of the problem in Saint Louis, the Chicago Health Department reported that the death rate from diphtheria had increased by more than 30 percent because many parents were still more frightened by the antiserum than the disease. Children died as a result. Again, a bad choice.

After these problems had been fixed, the choice to avoid the polio vaccine or sulfanilamide or diphtheria antiserum exposed children to unnecessary risks. Tragedies are inevitable. But this doesn't mean that people shouldn't trust that scientists or regulators can fix their mistakes.

#4: Once the relative risks are known, choose the lesser risk—even if it might seem counterintuitive.

But what if the problem hasn't been fixed?

One of the most common infections in the world is caused by dengue virus. Every year, this virus infects about four hundred million people. Initial symptoms include headache, nausea, vomiting, rash, fever, and muscle, bone, and joint pain. Most people recover in a week or so. Some, however, develop more severe symptoms when the virus causes leaky blood vessels. At this point, symptoms are far more frightening, including easy bruising, severe abdominal pain, low blood pressure, difficult or rapid breathing, bleeding from the mouth and nose, and blood in urine and stool. About five hundred thousand people every year are hospitalized with these symptoms. As the disease worsens, some go into shock and die; twenty thousand people in the world die every year from what is called dengue hemorrhagic shock syndrome.

Strangely, those who are most likely to die from this shock syndrome have previously been infected with the virus. In other words, it's the second dengue virus infection that is more likely to kill people, not the first.

How is this possible?

Researchers have identified four different strains of dengue virus. People infected with one dengue type won't be infected with that same type again. They're immune. However, people infected with one type who are later infected with a different type are more likely to go into shock and die than those who had never been infected with the virus before. The reason, which is complicated, is now understood.

People infected with one dengue type develop antibodies that prevent that type from attaching to cells and infecting them. However, at the same time that people are making these virus-neutralizing antibodies, they are also making virus-binding antibodies for the other three dengue types. Virus-binding antibodies don't neutralize dengue virus; rather, they facilitate its entrance into cells, making the second infection with a different dengue type worse. Much worse. In other words, the virus-binding antibodies, because they don't neutralize the virus, actually *increase* the chance that a subsequent infection with one of the other three types will be fatal.

Researchers at Sanofi Pasteur who made the first dengue vaccine, called Dengvaxia, knew about this problem. But they had a solution. Using a yellow fever vaccine as a backbone, they inserted the genes representing the surface proteins of dengue types 1, 2, 3, and 4. People inoculated with this vaccine could now develop virus-neutralizing antibodies against all four dengue types at the same time. These virus-neutralizing antibodies would then protect against any of the four strains likely to be encountered. Problem solved.

Unfortunately, for some young children, a devastating side effect was discovered. At the same time that Dengvaxia induced virus-neutralizing antibodies to all four dengue types, it also induced virus-binding antibodies to all four types. For some but not all children, there were more of these dangerous virus-binding antibodies and they lasted longer than the virus-neutralizing antibodies. As a consequence, some vaccinated children who had never been infected with dengue before vaccination suffered a severe and occasionally fatal dengue infection when exposed to the natural virus for the first time. In other words, after encountering natural dengue virus, some vaccinated children were worse off than children who had never received the vaccine. In the Philippines, where this vaccine was initially tested, fourteen vaccinated children died of dengue hemorrhagic shock syndrome when they were later naturally infected with the virus. Although a review by the World Health Organization concluded that these children might have died because the vaccine just didn't work, Rose Capeding, the physician in charge of the program, was indicted for murder. If Capeding had been convicted, she would have faced up to forty-eight years in prison.

So, now what? Again, there is no risk-free choice. At a meeting at the CDC in February 2020, researchers from the WHO showed that children whose parents had chosen not to give them the dengue vaccine were eighteen times more likely to be hospitalized and ten times more likely to die from dengue than children whose parents had chosen the vaccine. In other words, the dengue vaccine saved far more lives than it took.

The WHO scientists showed that, on balance, more children benefited from receiving the dengue vaccine than from not receiving it. Nonetheless, we are far more fearful of doing something that causes harm (such as getting a vaccine) than not doing something that results in harm (such as leaving ourselves unprotected). The sin of commission will always be viewed as greater than the sin of omission. But doing nothing *is* doing something. A choice not to get a vaccine is not a risk-free choice, just a choice to take a different and, in this case, far more serious risk.

Fortunately, the dengue vaccine problem appears to have been solved by another company, Takeda, that is now on the brink of licensing a dengue vaccine that doesn't appear to have the same problem as Dengvaxia.

#5: Some patients (and even some doctors) occasionally embrace the notion that when someone is sick or dying, any therapy is worth a shot. But if a medicine doesn't work, it can only do nothing or hurt.

In the middle of 2020, the Trump administration heralded hydroxychloroquine, an antimalarial drug, as a breakthrough treatment for COVID-19. Because COVID-19 was common, studies to determine whether hydroxychloroquine worked and was safe were done relatively quickly. It soon became clear that hydroxychloroquine didn't treat or prevent the disease. Worse, about one of every ten people who received the drug developed an abnormal heart rhythm, some fatally. Therefore, COVID-19 patients who had chosen to wait for studies were spared a drug that didn't work and was potentially harmful. Sometimes it makes sense to wait, especially if the answers to questions can come relatively quickly.

#6: Animal testing can be falsely reassuring.

Much of the early discussions about COVID-19 vaccine centered on successful studies in mice, rats, monkeys, hamsters, and ferrets showing that the vaccines worked and were safe. However, products cannot be claimed to be safe in people until they are tested in people.

Before Jesse Gelsinger agreed to participate in a gene-therapy program at the University of Pennsylvania, researchers had injected the disabled adenovirus vector containing the gene he needed into thousands of mice and hundreds of monkeys, all without a serious problem. Similar testing had been performed by French researchers prior to the retrovirus-vector-leukemia tragedy and by Sanofi researchers prior to the Dengvaxia tragedy.

After Jesse died, Katrine Bosley, CEO of a gene-therapy company in Cambridge, Massachusetts, said, "The genetic context of a mouse or a rat has nothing to do with human genetic context. You just can't know. You are taking a greater leap into the unknown with these kinds of experimental medicines. The FDA has learned lessons. The industry has learned lessons. And I think we are all seeking to be very careful in how we advance. Yet the risk never goes away. That's what it takes to make new medicines."

David Weiner, a vaccine researcher at the Wistar Institute in Philadelphia, sums it up best: "Mice lie and monkeys exaggerate."

#7: In the end, no matter how well-informed you are about a new technology, you're gambling. But you're gambling either way.

Here's another example of the difficulties about deciding whether to embrace a new technology.

More than one hundred thousand people in the United States and twenty million worldwide suffer from sickle cell disease, which is caused by an abnormality of hemoglobin. Hemoglobin is the protein in red blood cells that carries oxygen from the lungs to the rest of the body. In patients with sickle cell disease, the hemoglobin is abnormal, causing red blood cells to collapse and form a small sickle instead of a plump sphere. Typically, normal red blood cells live in the bloodstream for more than a hundred days. Red blood cells of patients with sickle cell disease, on the other hand, live for only about fifteen days. As a consequence, people with sickle cell disease require frequent blood transfusions.

One of the more common symptoms in patients with sickle cell disease is caused when red blood cells stick together and block small vessels. The result is severe, debilitating pain that requires hospitalization and treatment with opioids. As a consequence, many older sickle cell patients have become opioid addicts. In addition to these pain crises, sickle cell patients suffer strokes, frequent infections, and loss of vision. Most people with sickle cell disease die in their midforties. The only known cure is a bone-marrow transplant, with its attendant risks.

Enter CRISPR, a gene-editing technology. It is now possible to remove cells in the bone marrow of sickle cell patients and modify genes so that red blood cells no longer contain abnormal hemoglobin. Theoretically, this could end the constant pain, severe infections, endless series of blood transfusions, drug addiction, and limited life span. Indeed, on July 6, 2020, Victoria Gray, a thirty-four-year-old wife and mother from Forest, Mississippi, celebrated her one-year anniversary after having received CRISPR therapy for her sickle cell disease. Before the therapy, Gray averaged seven hospitalizations a year for blood transfusions. In the year following her treatment, she hadn't been hospitalized once. CRISPR therapy can now be considered for young children with this disease, allowing them to live longer, healthier lives. Or it might cause an unforeseen problem that reduces a forty-year life span to two years. The technology is simply too new to know.

ACKNOWLEDGMENTS

I want to thank T. J. Kelleher for his guidance in the construction of this book; Bonnie Offit, Emily Offit, Will Offit, and Sean O'Connor for their helpful suggestions; and Anne Marie Gottsdanker for her courage, wisdom, bravery, and friendship.

REFERENCES

Chapter 1: Louis Washkansky: Heart Transplants

Altman, L. "Christiaan Barnard, 78, Surgeon for First Heart Transplant, Dies." *New York Times*, September 3, 2001.

Associated Press. "James D. Hardy, 84, Dies; Paved Way for Transplants." *New York Times*, February 21, 2003.

Beecher, H. K. "Ethical Problems Created by the Hopelessly Unconscious Patient." *New England Journal of Medicine* 278 (1968): 1425–1430.

Brink, J. G., C. Barnard, and J. Hassoulas. "The First Human Heart Transplant and Further Advances in Cardiac Transplantation at Groote Schuur Hospital and the University of Cape Town." *Cardiovascular Journal of Africa* 20 (2009): 31–35.

Burch, M., and P. Aurora. "Current Status of Paediatric Heart, Lung, and Heart-Lung Transplantation." *Archives of Disease in Childhood* 89 (2004): 386–389.

Cooper, D. K. C. "A Brief History of Cross-Species Organ Transplantation." *Proceedings of the Baylor University Medical Center* 25 (2012): 49–57.

Cooper, D. K. C., B. Ekser, and A. J. Tector. "A Brief History of Clinical Xenotransplantation." *International Journal of Surgery* 23 (2015): 205–210.

Deschamps, J. Y., F. A. Roux, P. Saï, and E. Gouin. "History of Xenotransplantation." *Xenotransplantation* 12 (2005): 91–109.

DiBardino, D. J. "The History and Development of Cardiac Transplantation." *Texas Heart Institute Journal* 26 (1999): 198–205.

Diethelm, A. G. "Ethical Decisions in the History of Organ Transplantation." *Annals of Surgery* 211 (1990): 505–520.

Drazner, M. "Too Few Hearts to Go Around: How Science Can Solve the Organ Dilemma." UT Southwestern Medical Center, April 25, 2018. https://utswmed.org/medblog/reducing-wait-for-heart-transplants/.

Goila, A. K., and M. Pawar. "The Diagnosis of Brain Death." *Indian Journal of Critical Care Medicine* 13 (2009): 7–11.

Hardy, J. D., C. M. Chavez, F. D. Kurrus, et al. "Heart Transplantation in Man." *Journal of the American Medical Association* 188 (1964): 114–122.

Hoffman, N. *Heart Transplants.* Farmington Hills, MI: Lucent Books, 2003.

Kantrowitz, A., J. D. Haller, H. Joos, et al. "Transplantation of the Heart in an Infant and Adult." *American Journal of Cardiology* 22 (1968): 782–790.

Kittleson, M. M. "Recent Advances in Heart Transplantation." *F1000 Research* 7 (2018): 1008.

Linden, P. K. "History of Solid Organ Transplantation and Organ Donation." *Critical Care Clinics* 25 (2009): 165–184.

McRae, D. *Every Second Counts: The Race to Transplant the First Human Heart.* New York: G. P. Putnam's Sons, 2006.

Meine, T. J., and S. D. Russell. "A History of Orthotopic Heart Transplantation." *Cardiology in Review* 13 (2005): 190–196.

Mendeloff, E. N. "The History of Pediatric Heart and Lung Transplantation." *Pediatric Transplantation* 6 (2002): 270–279.

Mezrich, J. D. *When Death Becomes Life: Notes from a Transplant Surgeon.* New York: HarperCollins, 2019.

Morales, D. L. S., W. J. Dreyer, S. W. Denfield, et al. "Over Two Decades of Pediatric Heart Transplantation: How Has Survival Changed?" *Journal of Thoracic and Cardiovascular Surgery* 133 (2007): 632–639.

Morris, T. *The Matter of the Heart: A History of the Heart in Eleven Operations.* New York: St. Martin's Press, 2017.

Nathoo, A. *Hearts Exposed: Transplants and the Media in 1960s Britain*. London: Palgrave Macmillan, 2009.

Patterson, C., and K. B. Patterson. "The History of Heart Transplantation." *American Journal of the Medical Sciences* 314 (1997): 190–197.

Simpson, E. "Medawar's Legacy to Cellular Immunology and Clinical Transplantation: A Commentary on Billingham, Brent and Medawar (1956) 'Quantitative Studies on Human Tissue Transplantation Immunity. III. Actively Acquired Tolerance.'" *Philosophical Transactions of the Royal Society B: Biological Sciences*, April 19, 2015.

Singh, S. S. A., N. Banner, C. Berry, and N. Al-Attar. "Heart Transplantation: A History Lesson of Lazarus." *Vessel Plus* (2018). https://doi.org/10.20517/2574-1209.2018.28.

Stark, T. *Knife to the Heart: The Story of Transplant Surgery*. London: Macmillan, 1996.

Starzl, T. E. "History of Clinical Transplantation." *World Journal of Surgery* 24 (2000): 759–782.

Stolf, N. A. G. "History of Heart Transplantation: A Hard and Glorious Journey." *Brazilian Journal of Cardiovascular Surgery* 32 (2017): 423–427.

Thompson, T. *Hearts: Of Surgeons and Transplants, Miracles and Disasters Along the Cardiac Frontier*. New York: McCall Publishing Company, 1971.

Weiss, E. S., J. G. Allen, N. D. Patel, et al. "The Impact of Donor-Recipient Sex Matching on Survival After Orthotopic Heart Transplantation." *Circulation: Heart Failure* 2 (2009): 401–408.

Chapter 2: Ryan White: Blood Transfusions

Fox, J. P., C. Manson, H. A. Penna, and M. Pará. "Observations on the Occurrence of Icterus in Brazil Following Vaccination Against Yellow Fever." *American Journal of Hygiene* 36 (1942): 68–116.

George, R. *Nine Pints: A Journey Through the Money, Medicine, and Mysteries of Blood*. New York: Metropolitan, 2018.

Green, D. *Linked by Blood: Hemophilia and AIDS*. London: Academic Press, 2016.

Gupta, A. S. "Bio-Inspired Nanomedicine Strategies for Artificial Blood Components." *WIREs Nanomedicine and Nanobiotechnology* (2017). https://doi.org/10.1002/wnan.1464.

Hargett, M. V., H. W. Burruss, and A. Donovan. "Aqueous-Based Yellow Fever Vaccine." *Public Health Reports* 58 (1943): 505–512.

Harmening, D. M. *Modern Blood Banking and Transfusion Practices*. 7th ed. Philadelphia: F. A. Davis, 2019.

Lebrecht, N. *Genius and Anxiety: How Jews Changed the World, 1847–1947*. New York: Scribner, 2019.

Lederer, S. E. *Flesh and Blood: Organ Transplantation and Blood Transfusion in Twentieth-Century America*. Oxford: Oxford University Press, 2008.

Moore, P. *Blood and Justice: The 17th Century Parisian Doctor Who Made Blood Transfusion History*. Chichester, England: John Wiley & Sons, 2003.

Pemberton, S. *The Bleeding Disease: Hemophilia and the Unintended Consequences of Medical Progress*. Baltimore: Johns Hopkins University Press, 2011.

Pirnia, G. "How Charlie Chaplin Changed Paternity Laws in America." *Mental Floss*, April 16, 2015. www.mentalfloss.com/article/631 58/how-charlie-chaplin-changed-paternity-laws-america.

Sawyer, W. A., K. F. Meyer, M. D. Eaton, et al. "Jaundice in Army Personnel in the Western Region of the United States and Its Relation to Vaccination Against Yellow Fever." *American Journal of Hygiene* 40 (1944): 35–107.

Seeff, L. B., G. W. Beebe, J. H. Hoofnagle, et al. "A Serologic Follow-Up of the 1942 Epidemic of Post-Vaccination Hepatitis in the United States Army." *New England Journal of Medicine* 316 (1987): 965–970.

Squires, J. E. "Artificial Blood." *Science* 295 (2002): 1002–1005.

Starr, D. *Blood: An Epic History of Medicine and Commerce*. New York: Perennial, 2002.

Tolich, D. J., and K. McCoy. "Alternative to Blood Replacement in the Critically Ill." *Critical Care Nursing Clinics of North America* 29 (2017): 291–304.

Tucker, H. *Blood Work: A Tale of Medicine and Murder in the Scientific Revolution*. New York: W. W. Norton & Company, 2011.

Chapter 3: Hannah Greener: Anesthesia

Aptowicz, C. O. *Dr. Mütter's Marvels: A True Tale of Intrigue and Innovation at the Dawn of Modern Medicine*. New York: Avery, 2015.

Bunker, J. P., and C. M. Blumenfeld. "Liver Necrosis After Halothane Anesthesia: Cause or Coincidence?" *New England Journal of Medicine* 268 (1963): 531–534.

Burney, F. *Selected Letters and Journals*. 1st ed. Oxford: Oxford University Press, 1986.

Defalque, R., and A. J. Wright. "An Anesthetic Curiosity in New York (1875–1900): A Noted Surgeon Returns to 'Open Drop' Chloroform." *Anesthesiology* 88 (1998): 549–551.

Fenster, J. M. *Ether Day: The Strange Tale of America's Greatest Medical Discovery and the Haunted Men Who Made It*. New York: HarperCollins, 2001.

Gribben, M. "The Strange Death of William Rice." *Malefactor's Register*. Accessed December 23, 2019. www.malefactorsregister.com /wp/the-strange-death-of-william-rice.

Keys, T. E. *The History of Surgical Anesthesia*. Huntington, NY: Robert E. Krieger Publishing Company, 1978.

Macdonald, A. G. "A Short History of Fires and Explosions Caused by Anesthetic Agents." *British Journal of Anaesthesia* 72 (1994): 710–722.

Patrick, N. "The 'Murder Castle' Which Was Home to One of the World's First Serial Killers," *Vintage News*, August 23, 2016. www.thevintagenews.com/2016/08/23/prirority-h-h-holmes-one -first-documented-serial-killers-built-entire-hotel-chicago -designed-specifically-murder-became-known-murder-castle/.

Payne, J. P. "The Criminal Use of Chloroform." *Anaesthesia* 53 (1998): 685–690.

Robinson, V. *Victory Over Pain: A History of Anesthesia*. New York: Henry Schuman, 1946.

Shephard, D. *From Craft to Specialty: A Medical and Social History of Anesthesia and Its Changing Role in Health Care*. Thunder Bay, Canada: York Point Publishing, 2009.

Snow, S. J. *Blessed Days of Anaesthesia: How Anaesthetics Changed the World*. Oxford: Oxford University Press, 2008.

Sykes, K., and J. Bunker. *Anesthesia and the Practice of Medicine: Historical Perspectives*. London: Royal Society of Medicine Press, 2011.

Wawersik, J. "Die Geschichte der Chloroformnarkose" (History of chloroform anesthesia). *Anaesthesiologie und Reanimation* 22 (1997): 144–152.

Winters, R. W. *Accidental Medical Discoveries: How Tenacity and Pure Dumb Luck Changed the World*. New York: Skyhorse, 2016.

Chapter 4: "Jim": Biologicals

Blanchard, W. "1901—October 19-Nov 7, Tetanus Tainted (Horse Blood) Diphtheria Antitoxin, St. Louis, MO." Deadliest American Disasters and Large-Loss-of-Life Events. Accessed October 12, 2019. www.usdeadlyevents.com/1901-oct-19-nov-7-tetanus -tainted-horse-blood-diphtheria-antitoxin-st-louis-mo-13/.

Bren, L. "The Road to the Biotech Revolution—Highlights of 100 Years of Biologics Regulation." *FDA Consumer Magazine*, January–February 2006.

Coleman, T. S. "Early Developments in the Regulation of Biologics." *Food and Drug Law Journal* 71 (2016): 544–573.

DeHovitz, R. E. "The 1901 St. Louis Incident: The First Modern Medical Disaster." *Pediatrics* 133 (2014): 964–965.

Linton, D. S. *Emil von Behring: Infectious Disease, Immunology, Serum Therapy*. Philadelphia: American Philosophical Society, 2005.

Philadelphia Times. "Ten Children Die of Lockjaw." November 10, 1901.

St. Louis Post-Dispatch. "City Anti-Toxin Caused Deaths." October 30, 1901.

Tiwari, T. S. P., and M. Wharton. "Diphtheria Toxoid." In *Plotkin's Vaccines*, edited by S. A. Plotkin, W. A. Orenstein, P. A. Offit, and K. Edwards. 7th ed. London: Elsevier, 2015.

"Unjustifiable Distrust of Diphtheria Antitoxin." *Journal of the American Medical Association* 37 (1901): 1396–1397.

Willrich, M. *Pox: An American History*. New York: Penguin Press, 2011.

Chapter 5: Joan Marlar: Antibiotics

A. G. N. "The Elixir Sulfanilamide-Massengill." *Canadian Medical Association Journal* (1937): 590.

Akst, J. "The Elixir Tragedy, 1937." *Scientist Magazine*, June 1, 2013.

Ballentine, C. "Taste of Raspberries, Tastes of Death: The 1937 Elixir Sulfanilamide Incident." *FDA Consumer Magazine*, June 1981.

Duan, D. "Elixir Sulfanilamide—A Drug that Kills." LabRoots, January 24, 2018. www.labroots.com/trending/chemistry-and-physics/7892/elixir-sulfanilamide-drug-kills.

Gupta, A., and L. K. Waldhauser. "Adverse Drug Reactions from Birth to Early Childhood." *Pediatric Clinics of North America* 44 (1997): 79–92.

Haag, H. B., and A. M. Ambrose. "Studies on the Physiological Effect of Diethylene Glycol: II. Toxicity and Fate." *Journal of Pharmacology and Experimental Therapeutics* 59 (1937): 93–100.

Jarmusik, N. "1937—Elixir Sulfanilamide." *Compliance in Focus* (blog). IMARC, September 8, 2014. www.imarcresearch.com/blog/bid/354713/1937-Elixir-Sulfanilamide.

Lesch, J. E. *The First Miracle Drugs: How the Sulfa Drugs Transformed Medicine*. New York: Oxford University Press, 2007.

Martin, B. J. *Elixir: The American Tragedy of a Deadly Drug*. Lancaster, PA: Barkerry Press, 2014.

New York Times. "'Death Drug' Hunt Covered 15 States." November 25, 1937.

Paine, M. F. "Therapeutic Disasters That Hastened Safety Testing of New Drugs." *Clinical Pharmacology & Therapeutics* 101 (2017): 430–434.

Von Oettingen, W. F., and E. A. Jirouch. "The Pharmacology of Ethylene Glycol and Some of Its Derivatives in Relation to Their Chemical Constitution and Physical Chemical Properties." *Journal of Pharmacology and Experimental Therapeutics* 42 (1931): 355–372.

Wax, P. M. "Elixirs, Diluents, and the Passage of the 1938 Federal Food, Drug, and Cosmetic Act." *Annals of Internal Medicine* 122 (1995): 456–461.

West, J. G. "The Accidental Poison That Founded the Modern FDA." *The Atlantic*, January 16, 2018.

Young, D. "Documentary Examines Sulfanilamide Deaths of 1937." American Society of Health-System Pharmacists, December 5, 2003. www.scribd.com/document/7212321/Sulfa-Nil-Amide-Deaths-of-1937.

Young, J. H. "The 'Elixir Sulfanilamide' Disaster." *Emory University Quarterly* 14 (1958): 230–237.

Chapter 6: Anne Gottsdanker: Vaccines

Gottsdanker, Anne. Interview with author. Lancaster, CA, June 12, 2000.

Offit, P. A. *The Cutter Incident: How America's First Polio Vaccine Led to the Growing Vaccine Crisis*. New Haven, CT: Yale University Press, 2005.

Chapter 7: Clarence Dally: X-Rays

Assmus, A. "Early History of X Rays." *Beam Line* (Summer 1995): 10–24.

Babic, R. R., G. S. Babic, S. R. Babic, and N. R. Babic. "120 Years Since the Discovery of X-Rays." *Medicinski pregled* 69 (2016): 323–330.

Berman, B. *Zapped: From Infrared to X-Rays, the Curious History of Invisible Light*. London: Oneworld Publications, 2018.

Dunlap, O. E. "Editorial: Deleterious Effects of X-Rays on the Human Body." *Electrical Review*, August 12, 1896.

Fabbri, C. N. *From Anesthesia to X-Rays: Innovations and Discoveries That Changed Medicine Forever*. Santa Barbara, CA: Greenwood, 2017.

Frankel, R. I. "Centennial of Röntgen's Discovery of X-Rays." *Western Journal of Medicine* 164 (1996): 497–501.

Gagliardi, R. A. "Clarence Dally: An American Pioneer." *American Journal of Roentgenology* 157 (1991): 922.

Glasser, O. *Wilhelm Conrad Röntgen and the Early History of the Roentgen Rays*. San Francisco: Norman Publishing, 1993.

Gunderman, R. B. *X-Ray Vision: The Evolution of Medical Imaging and Its Human Significance*. Oxford: Oxford University Press, 2013.

Herzig, R. "In the Name of Science: Suffering, Sacrifice, and the Formation of American Roentgenology." *American Quarterly* 53 (2001): 563–589.

Hessenbruch, A. "A Brief History of X-Rays." *Endeavour* 26 (2002): 137–141.

History.com. "German Scientist Discovers X-Rays." November 24, 2009. Last Modified November 6, 2020.

Howell, J. D. "Early Clinical Use of the X-Ray." *Transactions of the American Clinical and Climatological Association* 127 (2016): 341–349.

Kemerink, G. J., J. M. A. van Engelshoven, K. J. Simon, et al. "Early X-Ray Workers: An Effort to Assess Their Numbers, Risk, and Most Common (Skin) Affliction." *Insights into Imaging* 7 (2016): 275–282.

Kemerink, M., T. J. Dierichs, J. Dierichs, et al. "The Application of X-Rays in Radiology: From Difficult and Dangerous to Simple and Safe." *American Journal of Roentgenology* 198 (2012): 754–759.

King, G. "Clarence Dally—The Man Who Gave Thomas Edison X-Ray Vision." *Smithsonian Magazine*, March 14, 2012.

Lavine, M. "The Early Clinical X-Ray in the United States: Patient Experiences and Public Perception." *Journal of the History of Medicine and Allied Sciences* 67 (2011): 587–625.

Linton, O. W. "History of Radiology." *Academic Radiology* 19 (2012): 1304.

Linton, O. W. "X-Rays Can Harm You and Others." *Academic Radiology* 19 (2012): 260.

McClafferty, C. K. *The Head Bone's Connected to the Neck Bone: The Weird, Wacky, and Wonderful X-Ray*. New York: Farrar, Straus and Giroux, 2001.

Meggitt, G. *Taming the Rays: A History of Radiation and Protection*. Lymm, UK: Pitchpole Books, 2018.

Mould, R. F. "The Early History of X-Ray Diagnosis with Emphasis on the Contributions of Physics, 1895–1915." *Physics in Medicine and Biology* 40 (1995): 1741–1787.

Mould, R. F. "Invited Review: Röntgen and the Discovery of X-Rays." *British Journal of Radiology* 68 (1995): 1145–1176.

Nolan, D. J. "100 Years of X Rays." *British Medical Journal* 310 (1995): 614–615.

Rahhal, N. "The Martyr Who Gave Us Radiology: Visceral Photos Show the Fatal Wounds Thomas Edison's Assistant Sustained Trying to Help Develop X-Rays." *Daily Mail*, June 28, 2019.

Thiel, K. *X-Rays*. New York: Cavendish Square, 2018.

Warren Record. "Death of Clarence M. Dally, Thomas Edison's Chief Engineer, from Exposure to Roentgen Rays." October 7, 1904.

Widder, J. "The Origins of Radiotherapy: Discovery of Biological Effects of X-Rays by Freund in 1897, Kienböck's Crucial Experiments in 1900, and Still It Is the Dose." *Radiotherapy and Oncology* 112 (2014): 150–152.

Wolbarst, A. B. *Looking Within: How X-Ray, CT, MRI, Ultrasound, and Other Medical Images Are Created and How They Help Physicians Save Lives*. Berkeley: University of California Press, 1999.

Part III: Serendipity

BBC. "Syrian Children's Death 'Caused by Vaccine Mix-Up.'" September 18, 2014. www.bbc.com/news/world-middle-east-29251329.

Nebehay, S. "Human Error Seen in Measles Vaccination Deaths in Syria: WHO." Reuters, September 19, 2014. www.reuters.com /article/us-syria-crisis-measles-idUSKBN0HE1B020140919.

Chapter 8: Eleven Unnamed Children: Chemotherapy

American Cancer Society Medical and Editorial Content Team. "Evolution of Cancer Treatments: Chemotherapy." American Cancer Society. Last modified June 12, 2014. www.cancer.org/cancer/cancer-basics/history-of-cancer/cancer-treatment-chemo.html.

Christakis, P. "The Birth of Chemotherapy at Yale." *Yale Journal of Biology and Medicine* 84 (2011): 169–172.

DeVita, V. T. "The History of Chemotherapy." WNPR. https://medicine.yale.edu/media-player/3791/.

DeVita, V. T., and E. Chu. "A History of Cancer Chemotherapy." *Cancer Research* 68 (2008): 8643–8653.

DeVita, V. T., and E. DeVita-Raeburn. *The Death of Cancer*. New York: Farrar, Straus and Giroux, 2015.

Farber, S. "Some Observations on the Effect of Folic Acid Antagonists on Acute Leukemia and Other Forms of Incurable Cancer." *Blood* 4 (1949): 160–167.

Farber, S., E. C. Cutler, J. W. Hawkins, et al. "The Action of Pteroylglutamic Conjugates on Man." *Science* 106 (1947): 619–621.

Farber, S., L. K. Diamond, R. D. Mercer, et al. "Temporary Remissions in Acute Leukemia in Children Produced by Folic Acid Antagonist, 4-Aminopteroyl-Glutamic Acid (Aminopterin)." *New England Journal of Medicine* 238 (1948): 787–793.

Freireich, E. J. "The History of Leukemia Therapy—A Personal Journey." *Clinical Lymphoma, Myeloma & Leukemia* 12 (2012): 386–392.

Freireich, E. J. "The Road to Cancer Control Goes Through Leukemia Research." *Current Oncology* 16 (2009): 1–2.

Freireich, E. J., and N. A. Lemak. *Milestones in Leukemia Research and Therapy*. Baltimore: Johns Hopkins University Press, 1991.

Freireich, E. J., P. H. Wiernik, and D. P. Steensma. "The Leukemias: A Half-Century of Discovery." *Journal of Clinical Oncology* 31 (2014): 3463–3469.

Goodman, L. S., M. M. Wintrobe, W. Dameshek, et al. "Nitrogen Mustard Therapy: Use of Methyl-Bis(Beta-Chloroethyl)amine Hydrochloride and Tris(Beta-Chloroethyl)amine Hydrochloride for Hodgkin's Disease, Lymphosarcoma, Leukemia and Certain

Allied and Miscellaneous Disorders." *Journal of the American Medical Association* 132 (1946): 126–132.

Hutchings, B. L., E. L. R. Stokstad, N. Bohonos, and N. H. Slobodkin. "Isolation of New Lactobacillus Casei Factor." *Science* 99 (1944): 371.

Kantarjian, H. M., M. J. Keating, and E. J. Freireich. "Toward the Potential Cure of Leukemias in the Next Decade." *Cancer* 124 (2018): 4301–4313.

Laszlo, J. *The Cure of Childhood Leukemias: Into the Age of Miracles.* New Brunswick, NJ: Rutgers University Press, 1995.

Leuchtenberger, C., R. Lewisohn, R. Leuchtenberger, and D. Laszlo, "'Folic Acid' a Tumor Inhibitor." *Proceedings of the Society for Experimental Biology and Medicine* 55 (1944): 204–205.

Leuchtenberger, R., C. Leuchtenberger, D. Laszlo, and R. Lewisohn. "The Influence of 'Folic Acid' on Spontaneous Breast Cancers in Mice." *Science* 101 (1945): 46.

Lewisohn, R., C. Leuchtenberger, R. Leuchtenberger, and J. C. Keresztesy. "The Influence of Liver L. Casei Factor on Spontaneous Breast Cancer in Mice." *Science* 104 (1946): 436–437.

Markel, H. "Home Run King Babe Ruth Helped Pioneer Modern Cancer Treatment." PBS *NewsHour*, August 15, 2014. www.pbs.org/newshour/health/august-16-1948-babe-ruth-americas-greatest-baseball-star-pioneer-modern-treatment-cancer-dies.

Mills, S., J. M. Stickney, and A. B. Hagedorn. "Observations on Acute Leukemia in Children Treated with 4-Aminopteroylglutamic Acid." *Pediatrics* 5 (1950): 52–56.

Morrison, W. B. "Cancer Chemotherapy: An Annotated History." *Journal of Veterinary Internal Medicine* 24 (2010): 1249–1262.

Mukherjee, S. *The Emperor of All Maladies: A Biography of Cancer.* New York: Scribner, 2010.

New York Times. "War Gases Tried in Cancer Therapy." October 6, 1946.

Pui, C-H, and W. E. Evans. "A 50-Year Journey to Cure Childhood Acute Lymphoblastic Leukemia." *Seminars in Hematology* (2013) 50: 185–196.

Spain, P. D., and N. Kadan-Lottick. "Observations of Unprecedented Remissions Following Novel Treatment for Acute Leukemia in

Children in 1948." *Journal of the Royal Society of Medicine* 105 (2012): 177–181.

Tivey, H. "The Natural History of Untreated Acute Leukemia." *Annals of the New York Academy of Sciences* 60 (1954): 322–358.

Zubrod, C. G. "Historic Milestones in Curative Chemotherapy." *Seminars in Oncology* 6 (1979): 490–505.

Chapter 9: Jesse Gelsinger: Gene Therapy

Allen, J. *How Gene Therapy Is Changing Society*. San Diego, CA: Reference Point Press, 2015.

Angier, N. "Gene Experiment to Reverse Inherited Disease Is Working." *New York Times*, April 1, 1994.

Blaese, R. M., K. W. Culver, A. D. Miller, et al. "T Lymphocyte-Directed Gene Therapy for ADA-SCID: Initial Trial Results After 4 Years." *Science* 270 (1995): 475–480.

Branca, M. A. "Gene Therapy: Cursed or Inching Towards Credibility?" *Nature Biotechnology* 23 (2005): 519–521.

Braun, C. J., K. Boztug, A. Paruzynski, et al. "Gene Therapy for Wiskott-Aldrich Syndrome—Long-Term Efficacy and Genotoxicity." *Science Translational Medicine* 6 (2014): 1–14.

Cavazzana-Calvo, M., S. Hacein-Bey, G. de Saint Basile, et al. "Gene Therapy of Human Severe Combined Immunodeficiency (SCID)-X1 Disease." *Science* 288 (2000): 669–672.

Cohen, J. "Cancer Therapy Returns to Original Target: HIV." *Science* 365 (2019): 530.

Corrigan-Curay, J., O. Cohen-Haguenauer, M. O'Reilly, et al. "Challenges in Vector and Trial Design Using Retroviral Vectors for Long-Term Gene Correction in Hematopoietic Stem Cell Gene Therapy." *Molecular Therapy* 20 (2012): 1084–1094.

Cross, R. "The Redemption of James Wilson, Gene Therapy Pioneer." *Chemical & Engineering News*, September 12, 2019.

Gao, G., Y. Yang, and J. M. Wilson. "Biology of Adenovirus Vectors with E1 and E4 Deletions for Liver-Directed Gene Therapy." *Journal of Virology* 70 (1996): 8934–8943.

Grossman, M., S. E. Raper, K. Kozarsky, et al. "Successful *Ex Vivo* Gene Therapy Directed to Liver in a Patient with Familial Hypercholesterolemia." *Nature Genetics* 6 (1994): 335–341.

High, K. A., and M. G. Roncarlo. "Gene Therapy." *New England Journal of Medicine* 381 (2019): 455–464.

Jenks, S. "Gene Therapy Death—'Everyone Has to Share in the Guilt.'" *Journal of the National Cancer Institute* 92 (2000): 98–100.

Lehrman, S. "Virus Treatment Questioned After Gene Therapy Death." *Nature* 401 (1999): 517–518.

Manno, C., G. F. Pierce, V. R. Arruda, et al. "Successful Transduction of Liver in Hemophilia by AAV-Factor IX and Limitations Imposed by the Host Immune Response." *Nature Medicine* 12 (2006): 342–347.

Marshall, E. "Gene Therapy Death Prompts Review of Adenovirus Vector." *Science* 286 (1999): 2244–2245.

Maude, S. L., T. W. Laetsch, J. Buechner, et al. "Tisagenlecleucel in Children and Young Adults with B-Cell Lymphoblastic Leukemia." *New England Journal of Medicine* 378 (2018): 439–448.

Maude, S. L., D. T. Teachey, S. R. Rheingold, et al. "Sustained Remissions with CD19-Specific Chimeric Antigen Receptor (CAR)-Modified T Cells in Children with Relapsed/Refractory ALL." *Journal of Clinical Oncology* 34 (2016): 3011.

Mingozzi, F., M. V. Maus, D. J. Hui, et al. "$CD8^+$ T-Cell Responses to Adeno-Associated Virus Capsid in Humans." *Nature Medicine* 13 (2007): 419–422.

Mukherjee, S. "New Blood." *New Yorker*, July 22, 2019.

Mullen, C. A., K. Snitzer, K. W. Culver, et al. "Molecular Analysis of T Lymphocyte-Directed Gene Therapy for Adenosine Deaminase Deficiency: Long-Term Expression In Vivo of Genes Introduced with a Retroviral Vector." *Human Gene Therapy* 10 (1996): 1123–1129.

Mulligan, R., and P. Berg. "Expression of a Bacterial Gene in Mammalian Cells." *Science* 209 (1980): 1422–1427.

Muul, L. M., L. M. Tuschong, S. L. Soenen, et al. "Persistence and Expression of the Adenosine Deaminase Gene for 12 Years and Immune Reaction to Gene Transfer Components: Long-Term Results of the First Clinical Gene Therapy Trial." *Blood* 101 (2003): 2563–2569.

Nelson, D., and R. Weiss. "Gene Researchers Admit Mistakes, Liability." *Washington Post*, February 15, 2000.

Noguchi, P. "Risks and Benefits of Gene Therapy." *New England Journal of Medicine* 348 (2003): 193–194.

Nunes, F. A., E. E. Furth, and J. M. Wilson. "Gene Transfer into the Liver of Nonhuman Primates with E1-Delected Recombinant Adenoviral Vectors: Safety of Readministration." *Human Gene Therapy* 10 (1999): 2515–2526.

Ott, M. G., M. Schmidt, K. Schwarzwaelder, et al. "Correction of X-Linked Chronic Granulomatous Disease by Gene Therapy, Augmented by Insertional Activation of MDS1-EVI1, PRDM16 or SETBP1." *Nature Medicine* 12 (2006): 401–409.

Penn Today. "Institute for Human Gene Therapy Responds to FDA." February 14, 2000.

Raper, S. E., N. Chirmule, F. S. Lee, et al. "Fatal Systemic Inflammatory Response Syndrome in an Ornithine Transcarbamylase Deficient Patient Following Adenoviral Gene Transfer." *Molecular Genetics and Metabolism* 80 (2003): 148–158.

Raper, S. E., Z. J. Haskal, X. Ye, et al. "Selective Gene Transfer into the Liver of Non-Human Primates with E1-Deleted, E2A-Defective, or E1-E4 Deleted Recombinant Adenoviruses." *Human Gene Therapy* 9 (1998): 671–679.

Raper, S. E., M. Yudkoff, N. Chirmule, et al. "A Pilot Study of In Vivo Liver-Directed Gene Transfer with an Adenoviral Vector in Partial Ornithine Transcarbamylase Deficiency." *Human Gene Therapy* 13 (2002): 163–175.

Rinde, M. "The Death of Jesse Gelsinger, 20 Years Later." *Distillations,* June 4, 2019.

Savulescu, J. "Harm, Ethics Committees and the Gene Therapy Death." *Journal of Medical Ethics,* June 1, 2001. http://dx.doi.org/10.1136/jme.27.3.148.

Schnell, M. A., Y. Zhang, J. Tazelaar, et al. "Activation of Innate Immunity in Nonhuman Primates Following Intraportal Administration of Adenoviral Vectors." *Molecular Therapy* 3 (2001): 708–722.

Shalala, D. "Protecting Research Subjects—What Must Be Done." *New England Journal of Medicine* 343 (2000): 808–810.

Sibbald, B. "Death but One Unintended Consequence of Gene-Therapy Trial." *Canadian Medical Association Journal* 164 (2001): 1612.

Somia, N., and I. M. Verma. "Gene Therapy: Trials and Tribulations."
   Nature Reviews Genetics 1 (2000): 91–99.

Steinbrook, R. "The Gelsinger Case." In *The Oxford Textbook of
   Clinical Research Ethics*, edited by E. J. Emanuel, C. C. Grady,
   R. A. Crouch, et al., 110–120. New York: Oxford University
   Press, 2008.

Stolberg, S. G. "The Biotech Death of Jesse Gelsinger." *New York
   Times Magazine*, November 28, 1999.

Varnavski, A. N., R. Calcedo, M. Bove, et al. "Evaluation of Toxicity
   from High-Dose Systemic Administration of Recombinant Adenovirus Vector in Vector-Naïve and Pre-Immunized Mice." *Gene
   Therapy* 12 (2005): 427–436.

Varnavski, A. N., Y. Zhang, M. Schnell, et al. "Preexisting Immunity to
   Adenovirus in Rhesus Monkeys Fails to Prevent Vector-Induced
   Toxicity." *Journal of Virology* 76 (2002): 5711–5719.

Weiss, R., and D. Nelson. "Teen Dies Undergoing Experimental Gene
   Therapy." *Washington Post*, September 29, 1999.

Wenner, M. "Tribulations of a Trial." *Scientific American* 301 (2009):
   14–15.

Wilson, J. M. "Lessons Learned from the Gene Therapy Trial for
   Ornithine Transcarbamylase Deficiency." *Molecular Genetics and
   Metabolism* 96 (2009): 151–157.

Wilson, J. M. "Risks of Gene Therapy Research." *Washington Post*,
   December 6, 1999.

Wilson, R. F. "The Death of Jesse Gelsinger: New Evidence of the
   Influence of Money and Prestige in Human Research." *American
   Journal of Law and Medicine* 295 (2010).

Yang, Y., H. C. Ertl, and J. M. Wilson. "MHC Class I-Restricted
   Cytotoxic T Lymphocytes to Viral Antigens Destroy Hepatocytes
   in Mice Infected with E1-Deleted Recombinant Adenoviruses."
   Immunity 1 (1994): 433–442.

Ye, X., G. P. Gao, C. Pabin, et al. "Evaluating the Potential of Germ
   Line Transmission After Intravenous Administration of Recombinant Adenovirus in the C3H Mouse." *Human Gene Therapy* 9
   (1998): 2135–2142.

Ye, X., M. B. Robinson, M. L. Batshaw, et al. "Prolonged Metabolic
   Correction in Adult Ornithine Transcarbamylase-Deficient Mice

with Adenoviral Vectors." *Journal of Biological Chemistry* 271 (1996): 3639–3646.

Ye, X., M. B. Robinson, C. Pabin, et al. "Adenovirus-Mediated In Vivo Gene Transfer Rapidly Protects Ornithine Transcarbamylase-Deficient Mice from an Ammonium Challenge." *Pediatric Research* 41 (1997): 527–534.

Zhang, Y., N. Chirmule, G. Gao, et al. "Acute Cytokine Response to Systemic Adenoviral Vectors in Mice Is Mediated by Dendritic Cells and Macrophages." *Molecular Therapy* 3 (2001): 697–707.

Zimmer, C. "Gene Therapy Emerges from Disgrace to Be the Next Big Thing, Again." *Wired*, August 13, 2013.

Epilogue: Living with Uncertainty

Arin, F. "Dengue Vaccine Fiasco Leads to Criminal Charges for Researcher in the Philippines." *Science*, April 24, 2019.

Biswal, S., H. Reynales, X. Saez-Llorens, et al. "Efficacy of a Tetravalent Dengue Vaccine in Healthy Children and Adolescents." *New England Journal of Medicine* 381 (2019): 2009–2019.

Couzin, J., and J. Kaiser. "As Gelsinger Case Ends, Gene Therapy Suffers Another Blow." *Science* 307 (2005): 1028.

Dunn, A. "'This Is a Cure.' St. Jude's Gene Therapy Succeeds in 'Bubble Boy' Disease Study." *BioPharma Dive*, April 17, 2019. www .biopharmadive.com/news/st-jude-gene-therapy-scid-bubble -boy-disease-cure/552869.

Echaluce, C. C. "Dengue Shock Syndrome Caused Most of 14 Deaths." *Manila Bulletin*, February 7, 2018.

Hacein-Bey-Abina, S., C. Von Kalle, M. Schmidt, et al. "LMO2-Associated Clonal T Cell Proliferation in Two Patients After Gene Therapy for SCID-X1." *Science* 302 (2003): 415–419.

Howe, S. J., M. R. Mansour, K. Schwarzwaelder, et al. "Insertional Mutagenesis Combined with Acquired Somatic Mutations Causes Leukemogenesis Following Gene Therapy of SCID-X1 Patients." *Journal of Clinical Investigation* 118 (2008): 3143–3150.

Mamcarz, E., S. Zhou, T. Lockey, et al. "Lentiviral Gene Therapy Combined with Low-Dose Busulfan in Infants with SCID-X1." *New England Journal of Medicine* 380 (2019): 1525–1534.

Nam, C-H, and T. H. Rabbitts. "The Role of LMO2 in Development and in T Cell Leukemia After Chromosomal Translocation or Retroviral Insertion." *Molecular Therapy* 13 (2006): 15–25.

Plater, R. "First Person Treated for Sickle Cell Disease with CRISPR Is Doing Well." *Healthline*, July 6, 2020.

Sridhar, S., A. Leudtke, E. Langevin, et al. "Effect of Dengue Serostatus on Dengue Vaccine Safety and Efficacy." *New England Journal of Medicine* 379 (2018): 327–340.

Thomas, S. J., and I-K Yoon. "A Review of Dengvaxia: Development to Deployment." *Human Vaccines & Immunotherapeutics* 15 (2019): 2295–2314.

INDEX

APRIL SAUL

Paul A. Offit, MD, is the director of the Vaccine Education Center at the Children's Hospital of Philadelphia, as well as the Maurice R. Hilleman professor of vaccinology and professor of pediatrics at the Perelman School of Medicine at the University of Pennsylvania.